Spinor Genera
in Characteristic 2

MEMOIRS
of the
American Mathematical Society

Number 907

Spinor Genera
in Characteristic 2

Yuanhua Wang
Fei Xu

July 2008 • Volume 194 • Number 907 (second of 4 numbers) • ISSN 0065-9266

American Mathematical Society
Providence, Rhode Island

2000 *Mathematics Subject Classification.*
Primary 11E08, 11E12; Secondary 20G25, 20G30, 20G35.

Library of Congress Cataloging-in-Publication Data

Wang, Yuanhua, 1976–
　　Spinor genera in characteristic 2 / Yuanhua Wang, Fei Xu.
　　　　p. cm. — (Memoirs of the American Mathematical Society, ISSN 0065-9266 ; v. 194, no. 907)
　　"Volume 194, number 907 (second of 4 numbers)."
　　Includes bibliographical references.
　　ISBN 978-0-8218-4166-2 (alk. paper)
　　1. Spinor analysis.　I. Xu, Fei, 1963–　II. Title.

QA433.W36　2008
515′.63—dc22
　　　　　　　　　　　　　　　　　　　　　　　　　　　　　　　　　　　　　2008008509

Memoirs of the American Mathematical Society

　　This journal is devoted entirely to research in pure and applied mathematics.

　　Subscription information. The 2008 subscription begins with volume 191 and consists of six mailings, each containing one or more numbers. Subscription prices for 2008 are US$675 list, US$540 institutional member. A late charge of 10% of the subscription price will be imposed on orders received from nonmembers after January 1 of the subscription year. Subscribers outside the United States and India must pay a postage surcharge of US$38; subscribers in India must pay a postage surcharge of US$43. Expedited delivery to destinations in North America US$53; elsewhere US$130. Each number may be ordered separately; *please specify number* when ordering an individual number. For prices and titles of recently released numbers, see the New Publications sections of the *Notices of the American Mathematical Society*.

　　Back number information. For back issues see the *AMS Catalog of Publications*.

　　Subscriptions and orders should be addressed to the American Mathematical Society, P. O. Box 845904, Boston, MA 02284-5904, USA. *All orders must be accompanied by payment.* Other correspondence should be addressed to 201 Charles Street, Providence, RI 02904-2294, USA.

　　Copying and reprinting. Individual readers of this publication, and nonprofit libraries acting for them, are permitted to make fair use of the material, such as to copy a chapter for use in teaching or research. Permission is granted to quote brief passages from this publication in reviews, provided the customary acknowledgment of the source is given.

　　Republication, systematic copying, or multiple reproduction of any material in this publication is permitted only under license from the American Mathematical Society. Requests for such permission should be addressed to the Acquisitions Department, American Mathematical Society, 201 Charles Street, Providence, Rhode Island 02904-2294, USA. Requests can also be made by e-mail to reprint-permission@ams.org.

Memoirs of the American Mathematical Society (ISSN 0065-9266) is published bimonthly (each volume consisting usually of more than one number) by the American Mathematical Society at 201 Charles Street, Providence, RI 02904-2294, USA. Periodicals postage paid at Providence, RI. Postmaster: Send address changes to Memoirs, American Mathematical Society, 201 Charles Street, Providence, RI 02904-2294, USA.

　　© 2008 by the American Mathematical Society. All rights reserved.
This publication is indexed in *Science Citation Index*®, *SciSearch*®, *Research Alert*®, *CompuMath Citation Index*®, *Current Contents*®/*Physical, Chemical & Earth Sciences*.
Printed in the United States of America.

　　♾ The paper used in this book is acid-free and falls within the guidelines
established to ensure permanence and durability.
Visit the AMS home page at http://www.ams.org/

10 9 8 7 6 5 4 3 2 1　　13 12 11 10 09 08

Contents

Preface   vii

Chapter 1.   Local Theory   1
   1.1.   Some lemmas   1
   1.2.   W-type   5
   1.3.   Generation of $O(L)$ 14
   1.4.   Computations of spinor norms, I   22
   1.5.   Computations of spinor norms, II   26
   1.6.   Group structure of $\theta(X(L/K))$ 37
   1.7.   Reduction formula for $\theta(X(L/K))$, I   38
   1.8.   Reduction formula for $\theta(X(L/K))$, II   44
   1.9.   Some computation via reduction formulae   52
   1.10.   Some notation   57
   1.11.   Closed forms for $\theta(X(L/K))$ 59

Chapter 2.   Global Theory   63
   2.0.   Introduction   63
   2.1.   Number of spinor genera in a genus   66
   2.2.   Representations of spinor genera, codimension ≥ 2 74
   2.3.   Representations of spinor genera, codimension 0   82

Bibliography   85

ABSTRACT. The purpose of this paper is to establish the spinor genus theory of quadratic forms over global function fields in characteristic 2. The first part of the paper is to compute the integral spinor norms and relative spinor norms. The second part of the paper gives a complete answer to the integral representations of one quadratic form by another with more than four variables over a global function field in characteristic 2.

In chapter 1, we first solve the generation of local integral orthogonal groups by introducing the lattices of W-type. Then the local integral spinor norms of all integral lattices are computed in terms of Jordan decompositions and the relative integral spinor norms between two integral lattices are proved to be a group over local fields in characteristic 2. The relative spinor norms are also determined in terms of the Jordan decompositions of two integral lattices as well.

In chapter 2, we prove that the number of spinor genera in a genus can be any power of 2 over a global function field in characteristic 2. Moreover, the local-global principle holds for the integral representations of quadratic forms with codimension bigger than 2. When the codimension is less than or equal to 2, a necessary and sufficient condition for determining if a given lattice is represented by a spinor genus is given. As an application, infinitely many examples with codimension 2 such that the local-global principle for the integral representations of quadratic forms fails are provided. For codimension 0 case, the number of spinor genera in a genus which represent a given lattice can be any proper power of 2.

2000 *Mathematics Subject Classification.* Primary 11E08, 11E12; Secondary 20G25, 20G30, 20G35.

Key words and phrases. integral lattice, integral orthogonal group, spinor norm, class, spinor genus, genus

The second author was supported by NSFC grant # 10325105 and # 10531060. Received by editor November 11,2004, and in the revised form November 24,2005.

Preface

The quadratic Diophantine equations have a quite long and rich history. In the middle of 17-th century, Fermat studied the integers which can be represented by certain binary quadratic forms such as x^2+y^2 and x^2+2y^2. In 19-th century, Gauss laid the foundations of binary quadratic forms by using quadratic fields. He developed the composition theory, genus theory and so on, which was in his famous book "Disquisitions Arithmeticae ".

In the beginning of last century, Hilbert proposed his famous 23 problems. Among these problems, he asked to study the arithmetic properties of quadratic forms in his 11-th problem (see [OM2]). Almost at the same time, the p-adic completions of number fields have been developed and the quadratic forms over such fields had been studied. The rational representations of one quadratic forms by another can be solved by the local-global principle (Hasse-Minkowski theorem or Hasse principle) as part of the achievement of the global classfield theory (see [OM1]). However, such local-global principle fails for the integral representations of one quadratic form by another in general. In the first half of last century, the various analytical methods had been developed for studying the integral representations of quadratic forms such as the circle method of Hardy-Littlewood, Hecke's classical modular form theory, Linnik's ergodic theory and Siegel's mass formulae [Si] and so on.

Another important progress of quadratic form theory is Witt's geometric language [Wi] over general fields of characteristic not 2 in the thirties of the last century. By using such language, O'Meara [OM] solved the integral representations of quadratic forms over non-dyadic local fields and 2-adic local fields and Riehm [Ri] gave partial results for general dyadic local fields. In the fifties of last century, Eichler [Ei] studied quadratic forms from the algebraic group point of view and introduced so called spinor genus theory which became one of the most powerful tools to study the integral representations of quadratic forms over global fields via the strong approximation theorem for spin groups. Such spinor genus theory was further developed by various effort in [Kn], [Kn1], [We], [SP], [HSX] and was eventually complete in [CX] and [Xu3] recently.

It is well-known that the theory of quadratic forms (or orthogonal groups) in characteristic 2 has quite different feature from that of characteristic not 2. Arf [Ar] established the general theory of quadratic forms in characteristic 2 which is the analogy of Witt's work [Wi]. Later on, the integral equivalence for quadratic forms over local fields in characteristic 2 was solved in [Sa]. Riehm [Ri1] gave a partial answer to the integral representations of quadratic forms over local fields in characteristic 2. Harder [Ha] established reduction theory for semi-simple algebraic groups over global function fields, which implies the finiteness of class number of a integral quadratic form over a global field of characteristic 2. The relative elementary proof of the finiteness of class number in characteristic 2 was given in [Co].

In this paper, we will establish the spinor genus theory over global fields of characteristic 2. By the Strong Approximation Theorem for spin groups ([Pr]), this theory gives a complete solution of the problem to determine whether an integral quadratic form is represented by another integral indefinite quadratic form in at least 4 variables.

In the first chapter, we study the local theory. Our main concern is to compute the integral spinor norms and integral relative spinor norms which are important arithmetic invariants in the global theory. In order to compute the integral spinor norms and integral relative spinor norms, one needs to find good generators of the integral orthogonal groups. Such generation problem of integral orthogonal groups in characteristic 2 was only solved for modular lattices in [Po]. Based on this result, the integral spinor norms for modular lattices was computed in [Xu].

The main idea of this chapter can be described as follows. We first introduce the so called lattices of W-type. This notion, defined in terms of the Jordan splitting, turns out to be independent of Jordan splitting (which is highly non-unique in characteristic 2). Then we show the integral spinor norm is the whole non-zero elements of the ground field by explicit construction of integral symmetries if the lattice is of W-type in §1.2 and the integral orthogonal group is generated by integral symmetries and Eichler transformations if the lattice is not of W-type in §1.3. Therefore, for the purpose of computing integral spinor norms, we solve the generation problem of integral orthogonal groups completely. By using this result, we complete the computation of integral spinor norms of all lattices in §1.4 and §1.5. Based on the results in the previous sections, we prove that integral relative spinor norms have the natural group structure in §1.6. The reduction formulae for computing such integral relative spinor norms are established in §1.7 and §1.8. In the last three sections, the closed forms for all integral relative spinor norms are given by using the reduction formulae.

It should be pointed out that there are some similarities between dyadic theory and characteristic 2 theory. Some part of the above analogue has been established in [Xu1], [HSX] over dyadic local fields. By using such results, Beli [Be] computed the integral spinor norms over dyadic local fields in term of the so called BONGS (base of norm generators) he introduced. Such computations over dyadic local fields should be much simpler and more conceptual if one can introduce the right W-type of lattices in dyadic theory.

In the second chapter, we study the global theory based on the local computation in the first chapter. We first prove the realization theorem for spinor genera which implies that the number of spinor genera in a genus can be any power of 2 in §2.1. Then we study integral representations by spinor genera with codimension > 2, 2 and 0 respectively. For codimension > 2 case, following from the standard argument, every spinor genus in a genus represents the given quadratic form if this given form is represented by this genus. For codimension 2, either every spinor genus or exactly half of them in a genus represents the given quadratic form if this given form is represented by this genus. We also determine exactly when the exceptional case occur and which half part the spinor genus belong to in §2.2. Finally we study codimension 0 case and prove the realization theorem for representations by spinor genera in this case. It is also given to determine when the given quadratic form is represented by the spinor genus in §2.3.

Yuanhua Wang and Fei Xu

CHAPTER 1

Local Theory

1.1. Some lemmas

Our basic notations will be standard if not explained (see [OM1], [Sa] and [Ri1]). In this chapter, $\bar{F}$ denotes a finite field of characteristic 2, $F = \bar{F}((\pi))$ the formal Laurent series field over $\bar{F}$ with uniformizer π and $\mathfrak{o} = \bar{F}[[\pi]]$ the ring of formal power series. Let $\mathfrak{p}$ be the unique non-zero prime ideal of $\mathfrak{o}$ and $\mathfrak{u}$ the group of units in $\mathfrak{o}$. Write 0 and ρ the fixed representatives of $\bar{F}/\wp(\bar{F})$ where the mapping

$$\wp(x) = x^2 + x \quad \text{for any } x \in F$$

is a homomorphism of the additive group of F. Let ord be the ordinal function in F.

A quadratic space V over F means a finite dimensional vector space V over F with a map $Q : V \to F$ such that

$$Q(ax) = a^2 Q(x) \quad \forall a \in F \quad \text{and} \quad \langle x, y \rangle = Q(x+y) + Q(x) + Q(y)$$

is bilinear. Therefore a quadratic space has a sympletic structure as well in characteristic 2. We always assume that all quadratic spaces are non-defective which means that

$$x = 0 \quad \Leftrightarrow \quad \langle x, V \rangle = 0.$$

For any $\alpha \in \dot{F}$, V^α means the quadratic space of the same vector space with the map Q^α scaling by α.

For a non-defective quadratic space V, there is a sympletic basis (see [Ar])

$$\{e_1, f_1; \cdots ; e_n, f_n\}$$

of V such that

$$\langle e_i, f_j \rangle = \delta_{ij}, \quad \langle e_i, e_j \rangle = 0 \quad \text{and} \quad \langle f_i, f_j \rangle = 0.$$

This implies that the dimension of a non-defective quadratic space is always even. One can define the Arf invariant $\Delta(V)$ of V (see [Ar]) as

$$\Delta(V) = \sum_{i=1}^{n} Q(e_i) Q(f_i) \quad \text{in } F/\wp(F)$$

which is well-defined and independent of the choice of the sympletic basis (see [Ar]). Let

$$Ord(\Delta(V)) = \max\{\ ord(x)\ :\ x \in \Delta(V)\}.$$

By [Sa, Lemma 1.1], one has $Ord(\Delta(V)) = \infty, 0$ or a negative odd integer.

A lattice L in V means a finitely generated $\mathfrak{o}$-module in V such that FL is a non-defective quadratic space. A vector $x \in L$ is called primitive if x can be extended as a basis of L. A lattice $\mathfrak{o}x + \mathfrak{o}y$ of rank two is denoted by $(a,b)_r$ if

$$Q(x) = a, \quad Q(y) = b \quad \text{and} \quad \langle x, y \rangle = \pi^r.$$

When $a = b = 0$, this binary lattice is denoted by $\pi^r \mathbf{H}$ and is called $\mathfrak{p}^r$-hyperbolic plane.

The scale $\mathfrak{s}(L)$ (resp. norm $\mathfrak{n}(L)$) of a lattice L is defined as the fractional ideal generated by $\langle x, y \rangle$ (resp. $Q(x)$ for all $x \in L$) for all $x, y \in L$. Denote

$$s(L) = ord_{\mathfrak{p}} \mathfrak{s}(L) \text{ and } u(L) = ord_{\mathfrak{p}} \mathfrak{n}(L).$$

The volume $\mathfrak{v}(L)$ of L is defined as $det(\langle x_i, x_j \rangle) \mathfrak{o}$ for any $\mathfrak{o}$-module basis $\{x_1, \cdots, x_t\}$ of L.

The norm group $\mathfrak{g}(L)$ is defined as $\mathfrak{o}^2$-module generated by $\{Q(x) : x \in L\}$ which is in the form $a\mathfrak{o}^2 + b\mathfrak{o}^2$ where $a\mathfrak{o} = \mathfrak{n}(L)$ and a (resp. b) is a norm (resp. base) generator of L (see [Sa]). It is clear that $\mathfrak{s}(L) \subseteq \mathfrak{g}(L) \subseteq \mathfrak{n}(L)$. It should be pointed out that the conventions concerning $\mathfrak{s}(L)$, $\mathfrak{n}(L)$ and $\mathfrak{g}(L)$ differ from those of [OM1] in characteristic not 2 since the relation between quadratic form and bilinear form is different. Let

$$v(L) = 2s(L) + Ord(\Delta(FL)) - u(L)$$

for any lattice L and $\pi^\infty = 0$.

Let

$$O(V) = \{\sigma \in GL(V) : Q(\sigma x) = Q(x), \forall x \in V\}$$

be the orthogonal group of V. One can construct some elements in $O(V)$ as follows. Let $x \in V$ with $Q(x) \neq 0$. Then

$$\tau_x : V \to V; \quad \tau_x y = y + \frac{\langle x, y \rangle}{Q(x)} x, \quad \forall y \in V.$$

It is clear that $\tau_x \in O(V)$ and $\tau_{ax} = \tau_x$ for any $a \in \dot{F}$. Such τ_x is called a symmetry along x. It is well-known that every element in $O(V)$ can be written as a product of symmetries and the parity of numbers of symmetries is uniquely determined (see [De]). The special orthogonal group $SO(V)$ is defined as the elements in $O(V)$ which can be written as a product of even number of symmetries. Let $O(L)$ be the stablizer of L with the action of $O(V)$ and $SO(L) = SO(V) \cap O(L)$ for a lattice L in V.

By using the Clifford algebras of V, one can construct the spin group $Spin(V)$ over F and a group homomorphism

$$\theta : SO(V) \to \dot{F}/\dot{F}^2$$

such that the following sequence is exact (see [Ch])

$$Spin(V) \to SO(V) \to \dot{F}/\dot{F}^2.$$

We call $\theta(\cdot)$ the spinor norm map.

Let $G(V)$ the group generated by $\{Q(x) \neq 0 : x \in V\}$ in $\dot{F}$. For any two lattices L and K in V and $FL = V$, we define

$$X(L/K) = \{\sigma \in SO(V) : K \subseteq \sigma L\}.$$

In this section, we prove two very useful lemmas which play important roles in next several sections.

DEFINITION 1.1.1. For any $\gamma \in \mathfrak{u}$, there are $\alpha(\gamma), \beta(\gamma) \in \mathfrak{u}$ and a unique

$$d(\gamma) \in \{2k - 1 : k \in \mathbb{N}\} \cup \{\infty\}$$

such that
(c1) $$\gamma = \alpha(\gamma)^2 + \beta(\gamma)^2 \pi^{d(\gamma)}.$$
We call $d(\gamma)$ the defect of γ.

REMARK 1.1.2. $d(\gamma) = \infty$ if and only if $\gamma \in \mathfrak{u}^2$. It is clear that the representation of (c1) is unique when $d(\gamma) < \infty$.

LEMMA 1.1.3. *For any $\lambda, \delta, \epsilon \in \mathfrak{u}$, there exist $\alpha, \beta \in \mathfrak{u}$ such that*
$$\lambda = \alpha^2 \delta + \beta^2 \epsilon \pi^{d(\lambda\delta)}.$$

PROOF. Write
$$\epsilon\delta^{-1} = \alpha_1^2 + \beta_1^2 \pi^{d(\epsilon\delta)} \quad \text{and} \quad \lambda\delta^{-1} = \alpha_2^2 + \beta_2^2 \pi^{d(\lambda\delta)}$$
where $\alpha_i, \beta_i \in \mathfrak{u}$ for $1 \leq i \leq 2$. The result follows easily when $d(\lambda\delta) = \infty$ or $d(\epsilon\delta) = \infty$. Otherwise
$$d(\lambda\delta) + d(\epsilon\delta) \equiv 0 \mod 2$$
and let
$$x = \alpha_2 + \alpha_1^{-1}\beta_1\beta_2 \pi^{(d(\lambda\delta)+d(\epsilon\delta))/2} \in \mathfrak{u} \quad \text{and} \quad y = \alpha_1^{-1}\beta_2 \in \mathfrak{u}.$$
Then $\lambda = x^2\delta + y^2\epsilon\pi^{d(\lambda\delta)}$. □

For any two-dimensional quadratic space V, one has
$$G(V) = \begin{cases} \dot{F} & \text{if } Ord(\Delta(V)) = \infty \\ N_{E/F}(\dot{E}) & \text{otherwise} \end{cases}$$
where $E = F(\zeta)$ and $\zeta^2 + \zeta + \Delta = 0$ with
$$\Delta \in \Delta(V) \quad \text{and} \quad ord(\Delta) = Ord(\Delta(V)).$$
It is clear that $[\dot{F} : G(V)] \leq 2$ by the local class field theory (see [We1]). Since $G(V)$ is an open subgroup of $\dot{F}$, one has $1 + \mathfrak{p}^n \subseteq G(V)$ for sufficient large n. The following lemma gives the smallest such integer.

LEMMA 1.1.4. *Suppose* $\dim(V) = 2$ *and* $Ord(\Delta(V)) < \infty$. *Then*
$$1 + \mathfrak{p}^d \subseteq G(V) \quad \text{if and only if} \quad d > -Ord(\Delta(V)).$$

PROOF. If $Ord(\Delta(V)) = 0$, then $\rho \in \Delta(V)$. Since $\zeta^2 + \zeta + \rho$ is irreducible over $\bar{F}$, E is the unramified quadratic extension over F. Therefore
$$G(V) = \mathfrak{u}\dot{F}^2 = (1 + \mathfrak{p})\dot{F}^2$$
by the local class field theory (see [We1]).

We can assume $s = Ord(\Delta(V))$ is a negative odd integer by [Sa, Lemma 1.1] and $\sigma = \Delta\pi^{-s}$.

If $d > -s$, there exists $\varphi \in \mathfrak{u}$ such that
$$\varphi^2\sigma\pi + (1 + \pi^{(-s+1)/2})\varphi + 1 + \lambda\pi^{d+s-1} = 0$$
for any $\lambda \in \mathfrak{u}$ by Hensel's lemma. Let
$$x = 1 + \pi^{(-s+1)/2} \quad \text{and} \quad y = \varphi\pi^{-s+1}.$$
Then
$$N_{E/F}(x + y\zeta) = x^2 + xy + y^2\Delta = 1 + \lambda\pi^d \quad \text{and} \quad 1 + \lambda\pi^d \in G(V).$$

Since $1 + \mathfrak{p}^d$ is generated by $1 + \lambda \pi^d$ for all $\lambda \in \mathfrak{u}$, one has
$$1 + \mathfrak{p}^d \subseteq G(V).$$

Conversely, one only needs to show that
$$1 + \mathfrak{p}^{-s} \not\subseteq G(V).$$

Since the map $f : \bar{F} \to \bar{F}$ defined by
$$f(x) = \bar{\sigma} x^2 + x \quad \forall x \in \bar{F} \quad \text{where} \quad \bar{\sigma} \equiv \sigma \mod \pi$$
is an additive homomorphism with the kernel $\{0, \bar{\sigma}^{-1}\}$, there exists $\bar{\lambda} \in \dot{\bar{F}}$ such that
$$0 \notin \{\bar{\sigma} x^2 + x + \bar{\lambda} : x \in \bar{F}\}.$$

Then there exists $\lambda \in \mathfrak{u}$ such that
$$\{\sigma x^2 + x + \lambda : x \in \mathfrak{o}\} \subseteq \mathfrak{u}.$$

We claim that
$$1 + \lambda \pi^{-s} \notin G(V).$$

Otherwise there exist $x, y \in F$ satisfying
$$x^2 + xy + y^2 \Delta = 1 + \lambda \pi^{-s}.$$

By the domination principle in [Ri1, Lemma 1.5], we get $ord(x) = 0$. Consider the quadratic defect of the above equation, we have that $ord(y) = -s$. Let $y = \pi^{-s} y_1$ with $y_1 \in \mathfrak{u}$. Then
$$(x+1)^2 + (x+1) y_1 \pi^{-s} = \pi^{-s}(\sigma y_1^2 + y_1 + \lambda).$$

Since $\sigma y_1^2 + y_1 + \lambda \in \mathfrak{u}$ and $-s$ is an odd integer, a contradiction is derived from the above equation. $\square$

1.2. W-type

Any two-dimensional lattice can be written as
$$L = \mathfrak{o}x + \mathfrak{o}y = (\delta\pi^u, \varepsilon\pi^v)_s$$
where $u = u(L)$, $v = v(L)$ as defined in §1.0 and $\delta, \varepsilon \in \mathfrak{u}$. Here x can be chosen as any vector in L such that $ord(Q(x)) = u$. It is clear that $u \leq s$. Furthermore one has
$$u \equiv v + 1 \mod 2 \quad \text{and} \quad v < 2s - u$$
when $Ord(\Delta(FL)) < 0$. One can also see that L is a hyperbolic plane if and only if $u = s$ and $v = \infty$.

For any primitive vector $x \in L$, $\tau_x \in O(L)$ if and only if
$$\langle x, L \rangle \subseteq Q(x)\mathfrak{o}.$$
Let $S(L)$ be the subgroup of $O(L)$ generated by symmetries in $O(L)$.

LEMMA 1.2.1. *Suppose*
$$L = (\delta_1\pi^{u_1}, \varepsilon_1\pi^{v_1})_{s_1} \perp (\delta_2\pi^{u_2}, \varepsilon_2\pi^{v_2})_{s_2}$$
with $u_1 + u_2 \equiv 1 \mod 2$. Then $\theta(SO(L)) = \dot{F}$ if one of the following conditions holds

1) $u_1 = s_1$ or $u_2 = s_2$;
2) $u_1 + u_2 \leq 2s_1 + 1$.

PROOF. By scaling we can assume that $s_1 = 0$ and $\delta_1 = 1$. Suppose $\{e_1, f_1, e_2, f_2\}$ is the corresponding basis of L and
$$L_1 = \mathfrak{o}e_1 + \mathfrak{o}f_1 \;, \quad L_2 = \mathfrak{o}e_2 + \mathfrak{o}f_2.$$
It is obvious that $\tau_{e_1} \in O(L)$ and $\tau_{e_2} \in O(L)$. Since
$$\theta(\tau_{e_1}\tau_{e_2}) = \delta_2\pi^{u_1+u_2}$$
and $u_1 + u_2 \equiv 1 \mod 2$, we have $\delta_2\pi\dot{F}^2 \in \theta(SO(L))$.

1) $u_1 = 0$ or $u_2 = s_2$.
By [Xu, Prop. 1], we have
$$\theta(SO(L_1)) = \mathfrak{u}\dot{F}^2 \quad \text{or} \quad \theta(SO(L_2)) = \mathfrak{u}\dot{F}^2.$$
So $\theta(SO(L)) = \dot{F}$.

2) $u_1 + u_2 \leq 1$.
We consider the following two cases.
i) $u_1 > u_2$. By Lemma 1.1.3, for any $\mu \in \mathfrak{u}$, there exist η and ξ in $\mathfrak{u}$ such that
$$\mu = \eta^2 + \pi^{d(\mu)}\delta_2\xi^2.$$
Let
$$r = d(\mu) + u_1 - u_2 \quad \text{and} \quad \gamma = \eta e_1 + \pi^{r/2}\xi e_2.$$
Then $\tau_\gamma \in O(L)$. Since $\tau_{e_1} \in O(L)$ and
$$\theta(\tau_{e_1}\tau_\gamma) = Q(e_1)Q(\gamma) = \pi^{2u_1}(\eta^2 + \pi^{d(\mu)}\delta_2\xi^2) = \pi^{2u_1}\mu,$$
we have
$$\mathfrak{u}\dot{F}^2 \subseteq \theta(SO(L)).$$

ii) $u_1 < u_2$. By Lemma 1.1.3, for any $\mu \in \mathfrak{u}$, there exist η and ξ in $\mathfrak{u}$ such that
$$\mu = \eta^2 + \pi^{d(\mu)} \delta_2^{-1} \xi^2.$$
Let
$$r = d(\mu) + u_2 - u_1 \quad \text{and} \quad \zeta = \xi \pi^{r/2} e_1 + \eta e_2.$$
Then
$$Q(\zeta) = \delta_2 \pi^{u_2} \mu.$$
Since $u_1 + u_2 \leq 1$, one has $r/2 \geq u_2$. Note
$$\langle \zeta, L \rangle = \mathfrak{p}^{\min\{r/2, s_2\}} \quad \text{and} \quad s_2 \geq u_2,$$
we have that $\tau_\zeta \in O(L)$. So
$$\tau_{e_2} \tau_\zeta \in SO(L) \quad \text{and} \quad \mu \dot{F}^2 \in \theta(SO(L)).$$
We have that $\mathfrak{u}\dot{F}^2 \subseteq \theta(SO(L))$ and the result follows. $\square$

LEMMA 1.2.2. *Suppose*
$$L = (\delta_1 \pi^{u_1}, \varepsilon_1 \pi^{v_1})_{s_1} \perp (\delta_2 \pi^{u_2}, \varepsilon_2 \pi^{v_2})_{s_2}$$
with $u_1 \equiv u_2 \mod 2$. *Then* $\theta(SO(L)) = \dot{F}$ *if one of the following conditions is satisfied*
 1) $u_1 = s_1$ *and* $u_2 < v_2 \leq s_2$;
 2) $u_2 = s_2$ *and* $u_1 < v_1 \leq s_1$;
 3) $u_2 + u_1 \leq 2s_1$ *and* $u_1 < v_1 \leq s_1$ *or* $u_2 < v_2 \leq s_2$;
 4) $u_2 + u_1 \leq 2s_1$ *and* $d(\delta_2 \delta_1) \leq \min\{s_1 - (u_2+u_1)/2, s_1 - u_1, s_2 - u_2\}$;
 5) $u_2 \leq u_1$ *and* $u_1 \equiv s_1 + 1 \mod 2$;
 6) $u_2 - u_1 \geq 2(s_2 - s_1)$ *and* $u_2 \equiv s_2 + 1 \mod 2$.

PROOF. By scaling we can assume that $s_1 = 0$ and $\delta_1 = 1$. Suppose that $\{e_1, f_1, e_2, f_2\}$ is the corresponding basis of L. Let
$$L_1 = \mathfrak{o} e_1 + \mathfrak{o} f_1 \quad \text{and} \quad L_2 = \mathfrak{o} e_2 + \mathfrak{o} f_2.$$

1) $u_1 = 0$ and $u_2 < v_2 \leq s_2$.
By [Xu, Prop. 1], one has $\theta(SO(L_1)) = \mathfrak{u}\dot{F}^2$. By Lemma 1.1.4 and [Xu, Prop. 3],
$$\mathfrak{u}\dot{F}^2 \not\subseteq G(FL_2) = \theta(SO(L_2)) \subseteq \theta(SO(L)).$$
So $\theta(SO(L)) = \dot{F}$.

2) $u_2 = s_2$ and $u_1 < v_1 \leq 0$.
The result is obtained from 1) by considering the dual lattice of L.

3) $u_1 + u_2 \leq 0$.
We assume that $u_1 < v_1 \leq 0$. Otherwise we consider the dual lattice of L. Therefore we have $v_1 \equiv u_1 + 1 \mod 2$.
 i) $u_2 < v_1$. For any $\lambda \in \mathfrak{u}$, there exist α and $\beta \in \mathfrak{u}$ such that
$$1 + \lambda \pi = \alpha^2 + \beta^2 \delta_2^{-1} \varepsilon_1 \pi$$
by Lemma 1.1.3. Let
$$x = \alpha \pi^{(v_1 - u_2 - 1)/2} \in \mathfrak{o} \quad \text{and} \quad \eta = \beta f_1 + x e_2 \in L.$$
Then
$$Q(\eta) = \delta_2 \pi^{v_1 - 1}(1 + \lambda \pi).$$

Since $v_1 \leq 0$, we have
$$\tau_\eta \in O(L).$$
Then $\mathfrak{u}\dot{F}^2 \subseteq \theta(SO(L))$.

Since $u_1 < v_1 \leq 0$, one gets
$$\mathfrak{u}\dot{F}^2 \not\subseteq G(FL_1) = \theta(SO(L_1)) \subseteq \theta(SO(L))$$
by [Xu, Prop. 3] and Lemma 1.1.4. The result is obtained.

ii) $u_2 > v_1$. For any $\lambda \in \mathfrak{u}$, by Lemma 1.1.3, there exist $\alpha, \beta \in \mathfrak{u}$ such that
$$1 + \lambda \pi^{u_2 - v_1} = \alpha^2 + \varepsilon_1^{-1} \delta_2 \beta^2 \pi^{u_2 - v_1}.$$
Let $\eta = \alpha f_1 + \beta e_2$. Then
$$\tau_\eta \in O(L) \quad \text{and} \quad \theta(\tau_\eta) = \varepsilon_1 \pi^{v_1}(1 + \lambda \pi^{u_2 - v_1}).$$
It is obvious that $\tau_{f_1} \in O(L)$. Therefore
$$(1 + \mathfrak{p}^{u_2 - v_1})\dot{F}^2 \subseteq \theta(SO(L)).$$
Since
$$v_1 \leq 0 \quad \text{and} \quad u_2 - v_1 \leq -(u_1 + v_1),$$
one gets
$$\theta(SO(L_1)) = G(FL_1) \quad \text{and} \quad (1 + \mathfrak{p}^{u_2 - v_1}) \not\subseteq G(FL_1)$$
by [Xu, Prop. 3] and Lemma 1.1.4. Therefore $\theta(SO(L)) = \dot{F}$.

4) $u_2 + u_1 \leq 0$ and $d(\delta_2) \leq \min\{-(u_2+u_1)/2, -u_1, s_2 - u_2\}$.
Let $d = d(\delta_2)$. By Lemma 1.1.3, there exist $x_0, y_0 \in \mathfrak{u}$ such that
$$\pi^d = x_0^2 + y_0^2 \delta_2.$$
i) $u_1 < u_2$. Let
$$\eta = x_0 \pi^{(u_2 - u_1)/2} e_1 + y_0 e_2 \in L.$$
Then
$$Q(\eta) = x_0^2 \pi^{u_2} + y_0^2 \delta_2 \pi^{u_2} = \pi^{u_2} \pi^d$$
and
$$\tau_\eta \in O(L).$$
Therefore $\pi \dot{F}^2 \in \theta(SO(L))$.

For any $\lambda \in \mathfrak{u}$, by Lemma 1.1.3, there exist $\alpha, \beta \in \mathfrak{u}$ such that
$$1 + \lambda \pi = \alpha^2 + \beta^2 \pi.$$
Let
$$x = \alpha \pi^{(d-1)/2} + x_0 \beta \quad \text{and} \quad y = \beta y_0.$$
Then
$$\eta = x \pi^{(u_2 - u_1)/2} e_1 + y e_2 \in L \quad \text{and} \quad Q(\eta) = \pi^{u_2 + d - 1}(1 + \lambda \pi)$$
and
$$ord_{\mathfrak{p}}(\langle \eta, L \rangle) = \min\{\ ord(x) + (u_2 - u_1)/2,\ s_2\ \} \geq u_2 + d - 1.$$
Therefore
$$\tau_\eta \in O(L) \quad \text{and} \quad \mathfrak{u}\dot{F}^2 \subseteq \theta(SO(L))$$
and the result follows.

ii) $u_2 \leq u_1$. There is a new splitting $L = L_1' \perp L_2'$ such that
$$u(L_1') > u_1 \quad \text{and} \quad u(L_2') = u(L) = u_2.$$
If $u(L_1') \equiv u_2 + 1 \mod 2$, the result follows from Lemma 1.2.1.

Otherwise one can repeat the above argument and can assume that
$$L_1' \cong \mathbf{H}.$$
Therefore one always has that $\theta(SO(L)) \supseteq u\dot{F}^2$.

Let
$$\eta = x_0 e_1 + y_0 \pi^{(u_1-u_2)/2} e_2 \in L.$$
Then
$$Q(\eta) = x_0^2 \pi^{u_1} + y_0^2 \delta_2 \pi^{u_1} = \pi^{u_1+d}.$$
Since $u_1 + d \leq 0$, one has $\tau_\eta \in O(L)$. Therefore $\pi \dot{F}^2 \in \theta(SO(L))$ and the result follows.

5) $u_2 \leq u_1$ and $u_1 \equiv 1 \mod 2$.

By the same argument as above ii) of 4), one only needs to consider that L is split by $\mathbf{H}$. Therefore
$$u\dot{F}^2 \subseteq \theta(SO(L)).$$
Since $u_1 \equiv 1 \mod 2$, one has $\pi \dot{F}^2 \in \theta(SO(L))$ and the result follows.

6) $u_2 - u_1 \geq 2s_2$ and $u_2 \equiv s_2 + 1 \mod 2$.

The result follows by considering the dual lattice of L and 5). $\square$

REMARK 1.2.3. Suppose $u_1 \equiv u_2 \mod 2$, $v_1 \geq s_1$ and $v_2 \geq s_2$.
If $u_2 \leq u_1$ and $d(\delta_2 \delta_1) > s_1 - u_1$, then L is split by $\pi^{s_1} \mathbf{H}$.
If $s_2 - s_1 \leq (u_2 - u_1)/2$ and $d(\delta_2 \delta_1) > s_2 - u_2$, then L is split by $\pi^{s_2} \mathbf{H}$.

DEFINITION 1.2.4. A Jordan splitting
$$L = L_1 \perp L_2 \perp \cdots \perp L_t$$
with a norm generator $\delta_i \pi^{u(L_i)}$ of L_i for $1 \leq i \leq t$ is called W-type if one of the following conditions is satisfied
  1) $\dim(L_i) \geq 4$ and $ord(b_i) \leq s(L_i)$ where b_i is a base generator of L_i for some $1 \leq i \leq t$;
  2) $u(L_i) + u(L_j) \equiv 1 \mod 2$ and $u(L_i) + u(L_j) \leq 2s(L_i) + 1$ for some $i < j$;
  3) $u(L_i) \equiv u(L_j) \mod 2$ for some $i < j$ and $u(L_i) + u(L_j) \leq 2s(L_i)$ and
    i) $u(L_i) < v(L_i) \leq s(L_i)$ or $u(L_j) < v(L_j) \leq s(L_j)$; or
    ii) $u(L_j) \leq u(L_i)$ and $u(L_i) \equiv s(L_i) + 1 \mod 2$; or
    iii) $u(L_j) - u(L_i) \geq 2(s(L_j) - s(L_i))$ and $u(L_j) \equiv s(L_j) + 1 \mod 2$; or
    iv) $d(\delta_j \delta_i) \leq \min\{s(L_j) - u(L_j), s(L_i) - u(L_i), s(L_i) - (u(L_j) + u(L_i))/2\}$.

A lattice L is called W-type if there is a Jordan splitting of L of W-type.

The W-type of lattices is intrinsic and independent of Jordan splittings of lattices by the following proposition.

PROPOSITION 1.2.5. *If a Jordan splitting of L is of W-type, then any Jordan splitting of L is of W-type.*

1.2. W-TYPE

PROOF. Suppose
$$L = L_1 \perp \cdots \perp L_t$$
is a Jordan splitting of W-type. Then one can decompose each Jordan component L_i into orthogonal sum of two dimension lattices
$$L_i = L_{i1} \perp \cdots \perp L_{in_i}$$
for all $1 \leq i \leq t$ such that the complete decomposition (See [Sa, §2])
$$L = L_{11} \perp \cdots \perp L_{1n_1} \perp \cdots \perp L_{i1} \perp \cdots \perp L_{in_i} \perp L_{t1} \perp \cdots \perp L_{tn_t}$$
satisfies 2) and 3) in Definition 1.2.4.

Conversely, suppose
$$L = L_1 \perp \cdots \perp L_t$$
is a complete decomposition in the sense of [Sa, §2] and satisfies 2) and 3) in Definition 1.2.4. One can verify that the corresponding Jordan splitting by putting the orthogonal components with the same scalar together is a Jordan splitting of W-type. Indeed one only needs to show that $L_i \perp L_j$ satisfies 1) in Definition 1.2.4 if L_i and L_j satisfies 3) ii) or iv) in Definition 1.2.4 for $s(L_i) = s(L_j)$.

For case 3) ii), one has $ord(b_i) \leq s(L_i)$ where b_i is a base generator of L_i. This implies that $L_i \perp L_j$ satisfies 1) in Definition 1.2.4.

For case 3) iv), one can assume that $u(L_i) \leq u(L_j)$. Let $e_i \in L_i$ and $e_j \in L_j$ such that
$$Q(e_i) = \delta_i \pi^{u(L_i)}, \ \ Q(e_j) = \delta_j \pi^{u(L_j)} \ \text{and} \ \delta_i = \xi^2 \delta_j + \eta^2 \pi^d$$
where $\xi, \eta \in \mathfrak{u}$ and $d = d(\delta_i \delta_j)$ is odd by Lemma 1.1.3. Then
$$ord(Q(\pi^{\frac{u(L_j)-u(L_i)}{2}} e_i + \xi e_j)) = u(L_j) + d \equiv u(L_i) + 1 \mod 2$$
and
$$ord(Q(\pi^{\frac{u(L_j)-u(L_i)}{2}} e_i + \xi e_j)) = u(L_j) + d \leq s(L_j) = s(L_i).$$
This implies that $L_i \perp L_j$ satisfies 1) in Definition 1.2.4.

Now we show that any complete decomposition of L does not satisfy 2) and 3) in Definition 1.2.4 if there is a complete decomposition of L does not satisfy 2) and 3) in Definition 1.2.4.

Let
$$L = L_1 \perp \cdots \perp L_t = L'_1 \perp \cdots \perp L'_t$$
be two complete decompositions of L and $\{e_i, f_i\}$ is the basis of L_i such that

(c2) $$ord(Q(e_i)) = u(L_i) \ \text{and} \ ord(Q(f_i)) = v(L_i)$$

for $1 \leq i \leq t$. By [Sa, Thm.2.3], the second complete decomposition can be obtained from the first one by applying a sequence of the elementary transformations introduced in [Sa,§2].

Suppose the first complete decomposition does not satisfy 2) and 3) in Definition 1.2.4. In our case, one only needs to verify that the new complete decomposition does not satisfy 2) and 3) in Definition 1.2.4 by applying the following four elementary transformations either.

$$\begin{cases} e'_k = e_k + ae_{k+1}, & f'_k = \xi f_k, & e'_{k+1} = \eta e_{k+1}, & f'_{k+1} = f_{k+1} + a\pi^{b_k} f_k \\ e'_k = e_k + af_{k+1}, & f'_k = \xi f_k, & e'_{k+1} = e_{k+1} + a\pi^{b_k} f_k, & f'_{k+1} = \eta f_{k+1} \\ e'_k = \xi e_k, & f'_k = f_k + ae_{k+1}, & e'_{k+1} = \eta e_{k+1}, & f'_{k+1} = f_{k+1} + a\pi^{b_k} e_k \\ e'_k = \xi e_k, & f'_k = f_k + af_{k+1}, & e'_{k+1} = e_{k+1} + a\pi^{b_k} e_k, & f'_{k+1} = \eta f_{k+1} \end{cases}$$

where

$$\xi, \eta \in \mathfrak{u}, \quad a \in \mathfrak{o} \quad \text{and} \quad b_k = s(L_{k+1}) - s(L_k).$$

It should be pointed out that the new basis $\{e'_k, f'_k\}$ for the new complete decomposition does not necessarily satisfy (c2). Here we will only check the first elementary transformation. The others can follow from the same argument or symmetry.

For simplicity, we will use u_i (resp. v_i, s_i) to denote $u(L_i)$ (resp. $v(L_i)$, $s(L_i)$) and u'_i (resp. v'_i, s'_i) to denote $u(L'_i)$ (resp. $v(L'_i)$, $s(L'_i)$).

Case I: $u_k \equiv u_{k+1} + 1 \mod 2$.

By Definition 1.2.4. 2), one has

$$u_k + u_{k+1} > 2s_k + 1.$$

Since

$$2ord(a) + 2b_k + v_k \geq 2(s_{k+1} - s_k) + u_k > s_{k+1},$$

we have

$$ord(Q(f'_{k+1})) > ord(Q(e'_{k+1}))$$

and

$$u'_k = u_k \quad \text{and} \quad u'_{k+1} = u_{k+1}$$

and

$$u_{k+1} < v_{k+1} \leq s_{k+1} \iff u'_{k+1} < v'_{k+1} \leq s'_{k+1}.$$

Therefore we only need to consider the case that

$$u_k \equiv u_i \mod 2 \quad \text{and} \quad u_k + u_i \leq 2\min\{s_k, s_i\}$$

for some $1 \leq i \leq t$ and to verify 3), iv) in Definition 1.2.4.

Since

$$\delta'_k = \delta_k + a^2 \delta_{k+1} \pi^{u_{k+1} - u_k} \quad \text{and} \quad u_{k+1} - u_k > s_k - u_k,$$

we have $d(\delta'_k \delta_i) = d(\delta_k \delta_i)$.

Case II: $u_k \equiv u_{k+1} \mod 2$ *and* $u_{k+1} \leq u_k$.

By Definition 1.2.4. 3), i) and iv), one has

$$u'_k \equiv u_k \mod 2 \quad \text{or} \quad u'_k = s_k.$$

Since $s_k \equiv u_k \mod 2$ by Definition 1.2.4. 3), ii), one always has

$$u'_k \equiv u_k \mod 2 \quad \text{and} \quad u'_k \geq u_{k+1}.$$

By Definition 1.2.4.3),i), we have

$$v'_k \geq v_k > s_k \quad \text{and} \quad u'_{k+1} = u_{k+1}$$

and

$$v'_{k+1} \geq \min\{v_{k+1}, v_k + b_k\} > s_{k+1}.$$

Suppose there exists $1 \leq i \leq t$ with $u_i \equiv u'_k + 1 \mod 2$ such that

$$u'_k + u_i \leq 2\min\{s_k, s_i\} + 1.$$

Since
$$u'_k \geq u_{k+1} \text{ and } s_{k+1} \geq s_k,$$
one has
$$u_{k+1} + u_i \leq 2\min\{s_{k+1}, s_i\} + 1$$
which contradicts Definition 1.2.4. 2).

Suppose there exists $1 \leq i \leq t$ with $u_i \equiv u'_k \mod 2$ such that
$$u'_k + u_i \leq 2\min\{s_k, s_i\}.$$
It is clear that one only needs to verify that $ii)$ with $i < k$ or $iii)$ with $i > k$ can not be true among Definition 1.2.4 3), $i) - iii)$. Suppose not.

For $ii)$, one has
$$u_{k+1} \leq u'_k \leq u_i \text{ and } u_i \equiv s_i + 1 \mod 2.$$
This implies the original complete splitting satisfies Definition 1.2.4. 3), $ii)$ for i and $k + 1$. A contradiction is derived.

For $iii)$, it is clear that $i > k+1$ and
$$u_i - u_{k+1} \geq u_i - u'_k \geq 2(s_i - s_k) \geq 2(s_i - s_{k+1}).$$
A contradiction is also derived as above.

Now we show that Definition 1.2.4. 3), $iv)$ can not be satisfied for the new complete splitting either. It is clear that one can assume that $u'_k < s_k$. Then
$$\delta'_k = \begin{cases} \delta_k + c^2 \delta_{k+1} \pi^{-l} & \text{if } l \leq 0 \\ \delta_k \pi^l + c^2 \delta_{k+1} & \text{otherwise} \end{cases}$$
where
$$a = c\pi^{ord(a)} \text{ and } l = u_k - u_{k+1} - 2ord(a).$$
Suppose there is $1 \leq i \leq t$ such that
$$u'_k \equiv u_i \mod 2 \text{ and } u'_k + u_i \leq 2\min\{s_k, s_i\}.$$
If $l \leq 0$, then
$$u'_k \geq u_k \geq u_{k+1}.$$
By Definition 1.2.4.3), $iv)$, we have
$$d(\delta_h \delta_i) > \min\{s_i - u_i, \ s_k - u'_k, \ \min\{s_k, s_i\} - \frac{u'_k + u_i}{2}\}$$
for $k \leq h \leq k+1$. Since
$$d(\delta'_k \delta_i) \geq \min\{d(\delta_k \delta_i), d(\delta_{k+1} \delta_i) - l\} \geq \min_{k \leq h \leq k+1}\{d(\delta_h \delta_i)\},$$
the result follows.

If $l > 0$, then
$$u'_k = u_{k+1} + 2ord(a) < u_k.$$
Since
$l + d(\delta_k \delta_i)$
$$\geq \begin{cases} l \geq \min\{s_i, s_k\} - \frac{u_i + u'_k}{2} & \text{if } u_i + u_k \geq 2\min\{s_i, s_k\} \\ l + \min\{s_i - u_i, \ s_k - u_k, \ \min\{s_i, s_k\} - \frac{u_k + u_i}{2}\} & \text{otherwise} \end{cases}$$
$$\geq \min\{s_i - u_i, \ s_k - u'_k, \ \min\{s_i, s_k\} - \frac{u'_k + u_i}{2}\}$$

and
$$d(\delta_{k+1}\delta_i) > \min\{s_i - u_i,\ s_k - u'_k,\ \min\{s_i, s_k\} - \frac{u'_k + u_i}{2}\}$$
by Definition 1.2.4.3),iv), one has
$$d(\delta'_k \delta_i) \geq \min\{l + d(\delta_k \delta_i),\ d(\delta_{k+1}\delta_i)\}$$
and the result follows.

Case III: $u_k \equiv u_{k+1} \mod 2$ *and* $u_k < u_{k+1}$.

It is clear that
$$ord(Q(a\pi^{b_k} f_k)) \geq 2(s_{k+1} - s_k) + v_k$$
$$\geq \begin{cases} s_{k+1} & \text{if } v_k \geq s_k \\ 2s_{k+1} - u_{k+1} - u_k + v_k \geq s_{k+1} & \text{otherwise} \end{cases}$$
by Definition 1.2.4.3), i). Moreover the equality holds only if
$$u_k = v_k = s_k = s_{k+1}$$
which is impossible by our assumption. So
$$ord(Q(f'_{k+1})) \geq \min\{v_{k+1}, s_{k+1}\} \geq u_{k+1}.$$
Therefore
$$u'_h = u_h \text{ for } k \leq h \leq k+1 \quad \text{and} \quad v'_k = v_k$$
and
$$u_{k+1} < v_{k+1} \leq s_{k+1} \Leftrightarrow u'_{k+1} < v'_{k+1} \leq s'_{k+1}.$$

Now one only needs to verify Definition 1.2.4.3), iv). Suppose there is $1 \leq i \leq t$ such that
$$u'_k \equiv u_i \mod 2 \quad \text{and} \quad u'_k + u_i \leq 2\min\{s_i, s_k\}.$$
Since
$$\delta'_k = \delta_k + a^2 \delta_{k+1} \pi^{u_{k+1} - u_k},$$
one has
$$d(\delta'_k \delta_i) \geq \min\{\ d(\delta_k \delta_i),\ d(\delta_{k+1}\delta_i) + u_{k+1} - u_k\}.$$
If $u_{k+1} + u_i \leq 2s_{k+1}$, then
$$d(\delta_{k+1}\delta_i) + u_{k+1} - u_k$$
$$> u_{k+1} - u_k + \min\{s_i - u_i,\ s_{k+1} - u_{k+1},\ \min\{s_{k+1}, s_i\} - \frac{u_i + u_{k+1}}{2}\}$$
$$\geq \min\{s_i - u_i,\ s_k - u'_k,\ \min\{s_i, s_k\} - \frac{u_i + u'_k}{2}\}$$
by Definition 1.2.4.3), iv).

Otherwise, we have
$$u_{k+1} - u_k > s_k - \frac{u'_k + u_i}{2} \geq \min\{s_i, s_k\} - \frac{u_i + u'_k}{2}$$
and the result follows. $\square$

COROLLARY 1.2.6. *If L is not of W-type, then any orthogonal component of L is not of W-type.*

PROOF. It follows from the proof of Prop.1.2.5. $\square$

PROPOSITION 1.2.7. *If L is of W-type, then $\theta(SO(L)) = \dot{F}$.*

PROOF. The proposition follows from [Xu, Prop. 2 and 3], Lemma 1.2.1 and Lemma 1.2.2. □

1.3. Generation of $O(L)$

Suppose i is a non-zero isotropic vector in V and $w \in V$ such that $\langle i, w \rangle = 0$. The Eichler transformation $E_w^i : V \mapsto V$ is defined as follows:
$$E_w^i x = x + \langle x, i \rangle w + \langle x, w \rangle i + Q(w)\langle x, i \rangle i \quad \text{for all} \quad x \in V.$$
It is easy to check that $E_w^i \in O(V)$ and
$$E_{w_1+w_2}^i = E_{w_1}^i E_{w_2}^i \quad \text{and} \quad E_w^i = \tau_{w+Q(w)i}\tau_w \quad \text{if} \quad Q(w) \neq 0.$$
If $Q(w) = 0$, there is $w' \in V$ such that
$$Q(w') = 0, \quad \langle w, w' \rangle \neq 0 \quad \text{and} \quad \langle i, w' \rangle = 0.$$
Let $a \in \dot{F}$ and $a \neq 1$. Then
$$E_w^i = E_{aw+w'}^i E_{(1-a)w+w'}^i.$$
So we always have $\theta(E_w^i) \in \dot{F}^2$ for any Eichler transformation E_w^i.

We denote $E(L)$ as the subgroup of $O(L)$ which is generated by $S(L)$ and those E_w^i in $O(L)$.

Let
$$L = L_1 \perp L_2 \perp \cdots \perp L_t$$
be a Jordan splitting of L.

PROPOSITION 1.3.1. *Suppose L is not of W-type and $\dim(L_1) = 2$. Assume*
$$u(L_1^\perp) - u(L_1) < 2(s(L_2) - s(L_1))$$
or
$$u(L_j) - u(L_1) = 2(s(L_2) - s(L_1))$$
for some $2 < j \leq t$ when
$$u(L_1) + s(L_2) < 2s(L_1).$$
If $\sigma L_1 \subseteq L$ for some $\sigma \in O(FL)$, then there is $\tau \in E(L)$ such that $\tau\sigma|_{L_1} = 1$.

PROOF. Write
$$L_1 = (\delta_1 \pi^{u_1}, \varepsilon_1 \pi^{v_1})_{s(L_1)} \quad \text{with the basis} \quad \{e_1, f_1\}$$
as explained in §1.2. By scaling we can assume that $s(L_1) = 0$ and $\delta_1 = 1$.

When $t = 1$, it has been done in [Po]. We assume $t > 1$.

Put $\sigma e_1 = ae_1 + bf_1 + z$ where $a, b \in \mathfrak{o}$ and $z \in L_1^\perp$. Since
$$\langle \sigma e_1, \sigma f_1 \rangle = \langle e_1, f_1 \rangle = 1,$$
we have that $a \in \mathfrak{u}$ or $b \in \mathfrak{u}$. It is clear that
$$Q(\sigma e_1 + e_1) = \langle \sigma e_1, e_1 \rangle = b.$$

Claim: there is $\varrho \in E(L)$ such that $\varrho \sigma e_1 = e_1$.

We can assume that $ord(b) > 0$. Otherwise
$$\tau_{\sigma e_1 + e_1} \in S(L) \quad \text{and} \quad \tau_{\sigma e_1 + e_1} \sigma e_1 = e_1.$$
Therefore one has $a \in \mathfrak{u}$.

Case I: $u_1 < u(L_1^\perp)$.

Then $a \equiv 1 \mod \pi$. It is clear that
$$\tau_{\sigma e_1 + e_1} \in S(L) \quad \text{if} \quad ord(b) = 1.$$
We can assume $ord(b) > 1$.

When L_1 is not a hyperbolic plane, one takes
$$\eta = \pi^{[-\frac{u_1}{2}]} e_1 + f_1 \in L.$$
Then
$$ord(Q(\eta)) \leq 0 \quad \text{and} \quad \tau_\eta \in S(L).$$
(i) $v_1 \geq 0$. Then
$$-1 \leq ord(Q(\eta)) \leq 0.$$
One can replace σe_1 by $\tau_\eta \sigma e_1$ and reduce to the above cases.

(ii) $v_1 < 0$. Then
$$ord(Q(\eta)) = v_1.$$
We can assume $ord(b) \leq -v_1$. Otherwise we can replace σe_1 by $\tau_\eta \sigma e_1$ if necessary.

Suppose $ord(b) < -v_1$. We consider
$$(a+1)^2 \pi^{u_1} = b^2 \varepsilon_1 \pi^{v_1} + ab + Q(z).$$
Since
$$ord(Q(z)) \geq u(L_1^\perp) \quad \text{and} \quad u(L_1^\perp) > -u_1 > -v_1$$
by Definition 1.2.4. 2) and 3) i), one has
$$2 ord(a+1) + u_1 = 2 ord(b) + v_1.$$
This contradicts the fact that $u_1 + v_1 \equiv 1 \mod 2$. Therefore we have
$$ord(b) = -v_1 \quad \text{and} \quad ord(a+1) > -v_1.$$
By Definition 1.2.4. 2) and 3) i), one has
$$-v_1 < -u_1 < u(L_i) \leq s(L_i)$$
for $i \geq 2$. Therefore we have $\tau_{\sigma e_1 + e_1} \in S(L)$.

When L_1 is a hyperbolic plane, then $Q(e_1) = 1$ and $Q(f_1) = 0$. Since
$$u(L_2) \geq u(L_1^\perp) > u_1 = 0$$
in this case, one has
$$s(L_2) \geq u(L_2) \geq 2$$
by Definition 1.2.4.2). One can assume that $ord(b) = 2$ when one replaces
$$\sigma e_1 \quad \text{by} \quad \tau_{e_1 + \pi f_1} \sigma e_1$$
if necessary.

If $ord(a+1) \geq 2$ or z is not a primitive vector in $L_1^\perp$, one can verify that
$$\tau_{\sigma e_1 + e_1} \in S(L)$$
and the result follows.

Otherwise, there are $a', b' \in \mathfrak{o}$ such that
$$E_z^{e_1 + f_1} \sigma e_1 = a' e_1 + b' f_1 + (a + 1 + b) z.$$
Since $u(L_1^\perp) > 0$, one has $E_z^{e_1 + f_1} \in O(L)$. It is clear that $a + 1 + b \in \mathfrak{p}$ and we reduce it to the above case.

Case II: *there exists some $2 \leq j \leq t$ such that $u(L_j) \leq u_1$.*

Then $u_1 \equiv u(L_j) \equiv 0 \mod 2$ and $v_1 \geq 0$ by Definition 1.2.4.2) and 3) i). Let
$$e_j \in L_j \quad \text{with} \quad ord(Q(e_j)) = u(L_j)$$
and
$$\eta = \begin{cases} f_1 + \pi^{-\frac{u(L_j)}{2}} e_j & \text{if } v_1 > 0 \\ f_1 & \text{if } v_1 = 0. \end{cases}$$
Then $\tau_\eta \in S(L)$ and there are $a' \in \mathfrak{o}$, $b' \in \mathfrak{u}$ and $z' \in L_1^\perp$ such that
$$\tau_\eta \sigma e_1 = a' e_1 + b' f_1 + z'.$$
One can replace σ by $\tau_\eta \sigma$ and reduce to the case that $ord(b) = 0$. Therefore the claim follows and we can assume that $\sigma e_1 = e_1$ in the following.

Claim: there is $\tau \in S(L)$ such that $\tau\sigma|_{L_1} = 1$.

Write
$$\sigma f_1 = x e_1 + y f_1 + w$$
where $x, y \in \mathfrak{o}$ and $w \in L_1^\perp$. Then
$$1 = Q(e_1, f_1) = Q(\sigma e_1, \sigma f_1) = y \quad \text{and} \quad Q(\sigma f_1 + f_1) = x.$$
Since one can replace σ by $\tau_{e_1}\sigma$ if necessary, one can assume that $ord(x) \leq -u_1$.

Case I: $u_1 + s(L_2) \geq 0$.

Since
$$ord(x) \leq -u_1 \leq s(L_2),$$
one has
$$\tau_{\sigma f_1 + f_1} \in S(L) \quad \text{and} \quad \tau_{\sigma f_1 + f_1} \sigma|_{L_1} = 1.$$

Case II: $u_1 + s(L_2) < 0$.

By the same argument for $ord(x) \leq s(L_2)$ as that in Case I, we only need to consider that $ord(x) > s(L_2)$. Let j be the maximal integer with $2 \leq j \leq t$ such that $u(L_j) = u(L_1^\perp)$.

Suppose $u(L_j) \leq u_1$. Then $u_j \equiv u_1 \mod 2$ by Definition 1.2.4.2). Let
$$e_j \in L_j \quad \text{and} \quad Q(e_j) = \delta_j \pi^{u(L_j)}.$$
By Definition 1.2.4.3)iv), there are α_j and $\beta_j \in \mathfrak{u}$ such that
$$\delta_j = \alpha_j^2 + \beta_j^2 \pi^{d(\delta_j)} \quad \text{and} \quad d(\delta_j) > -u_1.$$
Take an integer n such that
$$-\frac{u_1 + s(L_2)}{2} \leq n \leq \frac{-u_1}{2}.$$
Let
$$\lambda = (\alpha_j + \pi^n) e_1 + \pi^{\frac{u_1 - u(L_j)}{2}} e_j.$$
Then
$$ord(Q(\lambda)) = u_1 + 2n \leq 0 \quad \text{and} \quad \tau_\lambda \in O(L).$$
There are $x' \in \mathfrak{o}$ and $w' \in L_1^\perp$ such that
$$\tau_\lambda \sigma f_1 = x' e_1 + f_1 + w' \quad \text{and} \quad ord(x') = -u_1 - 2n \leq s(L_2).$$
By replacing σ by $\tau_\lambda \sigma$, one gets the result by the same argument for $ord(x) \leq s(L_2)$ as that in Case I.

Now we assume that $u_1 < u(L_j)$.

By the assumption, we have
$$u(L_j) + u_1 < 0 \quad \text{and} \quad -\frac{u(L_j) + u_1}{2} \leq \min\{-u_1, s(L_j) - u(L_j)\}.$$
By Definition 1.2.4.2) and 3)iv), one has
$$u_1 \equiv u(L_j) \mod 2 \quad \text{and} \quad d(\delta_j) > -\frac{u(L_j) + u_1}{2} \geq -\frac{u(L_2) + u_1}{2}.$$
If $u(L_j) - u_1 < 2s(L_2)$, then
$$\frac{u(L_j) + u_1}{2} - u_1 = \frac{u(L_j) - u_1}{2} < s(L_2) < -u_1.$$
Therefore
$$u_1 + u(L_j) \leq -4.$$
If $u(L_j) - u_1 = 2s(L_2)$ and $j > 2$, then one has
$$u(L_2) \equiv u_j \equiv u_1 \equiv s(L_2) \mod 2$$
by Definition 1.2.4.2) and 3)ii). So
$$\frac{u(L_j) + u_1}{4} = \frac{s(L_2) + u_1}{2} \in \mathbb{Z}.$$
Therefore $u(L_j) + u_1 \leq -4$.

Let
$$\delta_j = \alpha_j^2 + \beta_j^2 \pi^{d(\delta_j)} \quad \text{with} \quad \alpha_j, \beta_j \in \mathfrak{u}$$
and
$$\alpha = (\alpha_j + \pi^{[-\frac{u(L_j)+u_1}{4}]}) \pi^{\frac{u(L_j)-u_1}{2}}.$$
Take $\lambda = \alpha e_1 + e_j$. Then
$$ord(Q(\lambda)) = u(L_j) + 2[-\frac{u(L_j) + u_1}{4}] \leq \frac{u(L_j) - u_1}{2}$$
and $\tau_\lambda \in S(L)$. There are $x' \in \mathfrak{o}$ and $w' \in L_1^\perp$ such that
$$\tau_\lambda \sigma f_1 = x' e_1 + f_1 + w'$$
and
$$ord(x') = -u_1 - 2[-\frac{u(L_j) + u_1}{4}] \leq s(L_2).$$
By considering $\tau_\lambda \sigma$ instead of σ, one obtains the result by the same argument for $ord(x) \leq s(L_2)$ as that in Case I. □

COROLLARY 1.3.2. *Suppose L is not of W-type and $\dim(L_1) = 2$. Let*
$$e_1 \in L_1 \quad \text{with} \quad ord(Q(e_1)) = u(L_1).$$
Then there is $\tau \in E(L)$ such that $\tau \sigma e_1 = e_1$ for any $\sigma \in O(FL)$ with $\sigma e_1 \in L$. In particularly, if L_1 is not a hyperbolic plane, there exists $\tau \in S(L)$ such that $\tau \sigma e_1 = e_1$.

PROOF. It follows from the proof of Proposition 1.3.1. □

REMARK 1.3.3. The condition in Proposition 1.3.1

"$u(L_1^\perp) - u(L_1) < 2(s(L_2) - s(L_1))$ or $u(L_j) - u(L_1) = 2(s(L_2) - s(L_1))$

for some $2 < j \leq t$ when

$$u(L_1) + s(L_2) < 2s(L_1)"$$

can not be improved as

"$u(L_1^\perp) - u(L_1) \leq 2(s(L_2) - s(L_1))$

when

$$u(L_1) + s(L_2) < 2s(L_1)."$$

by the following example.

Let

$$L = L_1 \perp L_2 = L_1' \perp L_2'$$

where

$$L_1 = (\mathfrak{o}e_1 + \mathfrak{o}f_1) \cong (\pi^{u_1}, 0)_0$$

and

$$L_2 = (\mathfrak{o}e_2 + \mathfrak{o}f_2) \cong (\pi^{u_1+2s_2}, 0)_{s_2}$$

and

$$L_1' = (\mathfrak{o}e_1 + \mathfrak{o}(f_1 + f_2)) \cong (\pi^{u_1}, 0)_0$$

and

$$L_2' = (\mathfrak{o}(e_2 + \pi^{s_2}e_1) + \mathfrak{o}f_2) \cong (\pi^{s_2}, 0)_{s_2}$$

with $u_1 + s_2 < 0$. It is clear that there is $\sigma \in O(FL)$ such that $\sigma L_1 = L_1'$. If Proposition 1.3.1 is still true under the improved conditions, then there is $\tau \in O(L)$ such that $\tau\sigma|_{L_1} = 1$. So $\tau L_1' = L_1$. Therefore

$$L = L_1 \perp L_2 = \tau L = \tau L_1' \perp \tau L_2' = L_1 \perp \tau L_2'.$$

This implies that $\tau L_2' = L_2$ which is impossible.

It should be pointed out that Proposition 1.3.1 gives the Witt-type cancellation theorem for lattices. It is explained that the conditions in Proposition 1.3.1 can not be improved in Remark 1.3.3 unconditionally. However the following proposition shows that one can extend such result which we need for establishing the reduction formulae of relative spinor norms with some extra conditions.

PROPOSITION 1.3.4. *Suppose L is not of W-type, $\dim(L_i) = 2$ for $1 \leq i \leq 2$ and*

$$u(L_2) = u(L_1) + 2s(L_2) - 2s(L_1).$$

Let

$$e_2 \in L_2 \quad with \quad ord(Q(e_2)) = u(L_2).$$

If $\sigma L_1 \subseteq L$ and $\sigma e_2 \in L$ for some $\sigma \in O(FL)$, then there is $\tau \in S(L)$ such that

$$\tau\sigma|_{L_1} = 1.$$

PROOF. Let
$$L_i = \mathfrak{o}e_i + \mathfrak{o}f_i \cong (\delta_i \pi^{u_i}, \varepsilon_i \pi^{v_i})_{s(L_i)}$$
as explained in §1.2 for $i = 1, 2$. By scaling, one can assume $s(L_1) = 0$ and $\delta_1 = 1$.

Suppose $u(L_2) = s(L_2)$. Then $u(L_1) + s(L_2) = 0$ and the result follows from Corollary 1.3.2 and the same argument of Case I of the second claim of Proposition 1.3.1. Therefore we can assume that $u(L_2) < s(L_2)$. It is clear that L_1 can not be a hyperbolic plane by our assumption. By Corollary 1.3.2, we can assume that $\sigma e_1 = e_1$. Since
$$\langle \sigma e_1, \sigma f_1 \rangle = \langle e_1, f_1 \rangle = 1 \quad \text{and} \quad \langle \sigma e_1, \sigma e_2 \rangle = \langle e_1, e_2 \rangle = 0,$$
there are $x_i, y_i \in \mathfrak{o}$ for $1 \leq i \leq 4$ and $w, z \in L_3 \perp \cdots \perp L_t$ such that
$$\sigma f_1 = x_1 e_1 + f_1 + x_2 e_2 + y_2 f_2 + w$$
and
$$\sigma e_2 = x_3 e_1 + x_4 e_2 + y_4 f_2 + z.$$
By the same argument of the second claim in Proposition 1.3.1, one can assume that
$$s(L_2) < ord(x_1) \leq -u(L_1) \quad \text{and} \quad u(L_2) < u(L_j)$$
for $2 < j \leq t$.

Furthermore, we can assume that $y_2 \in \mathfrak{u}$. Otherwise, $y_2 \in \mathfrak{p}$. Since
$$u(L_2) \equiv u(L_1) \equiv s(L_2) \mod 2$$
and
$$d(\delta_2) > -\frac{u(L_1) + u(L_2)}{2}$$
by Definition 1.2.4.3)iii) and iv), one has
$$\delta_2 = \alpha^2 + \beta^2 \pi^{d(\delta_2)} \quad \text{with} \quad \alpha, \beta \in \mathfrak{u}$$
and
$$\lambda = (\alpha + \pi^{-\frac{u(L_1)+s(L_2)}{2}})\pi^{s(L_2)} e_1 + e_2 \in L.$$
Then
$$ord(Q(\lambda)) = s(L_2) \quad \text{and} \quad \langle \lambda, L \rangle = \mathfrak{p}^{s(L_2)}.$$
Therefore $\tau_\lambda \in S(L)$ and $\tau_\lambda(e_1) = e_1$. Replacing σ by $\tau_\lambda \sigma$, one obtains
$$ord(x_1) = s(L_2)$$
for the new x_1. The result follows from the same argument of the second claim in Proposition 1.3.1.

Since
$$Q(\sigma f_1) = Q(f_1)$$
and
$$u(L_2) < \min\{2ord(x_1) + u(L_1), s(L_2), ord(Q(w))\},$$
one has $x_2 \in \mathfrak{p}$. Since
$$u(L_2) = ord(Q(\sigma e_2)) < u(L_j)$$
for $3 \leq j \leq t$, one has
$$ord(x_3) \geq s(L_2).$$
Furthermore
$$ord(x_3) = s(L_2) \quad \text{or} \quad ord(x_4) = 0.$$

Since
$$\langle \sigma f_1, \sigma e_2 \rangle = x_3 + (x_4 y_2 + x_2 y_4)\pi^{s(L_2)} + \langle w, z \rangle = 0$$
and
$$x_2 \in \mathfrak{p} \quad \text{and} \quad y_2 \in \mathfrak{u},$$
one has
$$ord(x_3) - s(L_2) = ord(x_4) = 0 \quad \text{and} \quad x_3 \pi^{-s(L_2)} \equiv x_4 y_2 \mod \pi.$$
Suppose $y_4 \in \mathfrak{u}$. Then
$$Q(\sigma e_2 + e_2) = y_4 \pi^{s(L_2)} \quad \text{and} \quad \tau_{\sigma e_2 + e_2} \in S(L).$$
There are $x_1' \in \mathfrak{o}$ and $w' \in L_1^\perp$ such that
$$\tau_{\sigma e_2 + e_2} \sigma f_1 = x_1' e_1 + f_1 + w'$$
where
$$x_1' = x_1 + (x_3 + x_2 y_4 \pi^{s(L_2)} + y_2(x_4 + 1)\pi^{s(L_2)} + \langle z, w \rangle) x_3 y_4^{-1} \pi^{-s(L_2)}.$$
Then
$$ord(x_1') = ord(x_3 + y_2(x_4 + 1)\pi^{s(L_2)}) = ord(y_2) + s(L_2) = s(L_2).$$
It is clear that
$$\tau_{\sigma e_2 + e_2} \sigma e_1 = e_1.$$
One can replace
$$\sigma \quad \text{by} \quad \tau_{\sigma e_2 + e_2} \sigma$$
and the result follows.

Suppose $y_4 \in \mathfrak{p}$. By Definition 1.2.4.3)i) and iii), one has
$$s(L_2) \equiv u(L_2) \mod 2 \quad \text{and} \quad v_2 > s(L_2).$$
Let
$$\lambda = \pi^{(s(L_2) - u(L_2))/2} e_2 + f_2.$$
Then
$$ord(Q(\lambda)) = s(L_2), \quad \tau_\lambda \in S(L) \quad \text{and} \quad \tau_\lambda|_{L_1} = 1.$$
There are x_4' and y_4' in $\mathfrak{u}$ such that
$$\tau_\lambda \sigma e_2 = x_3 e_1 + x_4' e_2 + y_4' f_2 + z.$$
Replacing σ by $\tau_\lambda \sigma$, one reduces it to the above case. $\square$

COROLLARY 1.3.5. Suppose L is not of W-type, then $O(L) = E(L)$.

PROOF. Let
$$L = L_1 \perp \cdots \perp L_t$$
be the Jordan splitting of L. We prove the result by induction on $\dim(L)$.
If $\dim(L_i) \geq 4$, then
$$L_i = L_i' \perp H_i \quad \text{where} \quad \dim(L_i') \leq 2$$
and H_i is the sum of hyperbolic planes by Definition 1.2.4.1) for $1 \leq i \leq t$. By Corollary 1.2.6, we have
$$L' = L_1' \perp L_2' \perp \cdots \perp L_t'$$
is not of W-type. Therefore we can assume that $\dim(L_i) = 2$ and L_i is not a hyperbolic plane for $1 \leq i \leq n$ by [Po, 4.4].

By Proposition 1.3.1, we only need to consider that

$$u(L_1) + s(L_2) < 2s(L_1) \quad \text{and} \quad u(L_2) - u(L_1) \geq 2(s(L_2) - s(L_1)).$$

By Definition 1.2.4.2), one has $u(L_1) \equiv u(L_2) \mod 2$. There is a new Jordan splitting

$$L_1 \perp L_2 = K_1 \perp K_2$$

such that

$$u(K_2) > u(L_2),$$

which also satisfies the above condition. By repeating this argument, one can assume that K_2 is a hyperbolic plane. By Corollary 1.2.6, we have that

$$K_1 \perp L_3 \perp \cdots \perp L_t$$

is not of W-type. The result follows from [Po, 4.4] and induction. □

1.4. Computations of spinor norms, I

In the following two sections, we will compute $\theta(SO(L))$ for any lattice L. By Proposition 1.2.7, we can assume that L is not of W-type. Let

$$L = L_1 \perp L_2 \perp \cdots \perp L_t$$

be a Jordan splitting of L. Since L is not of W-type, one can write

$$L_i = (\mathfrak{o} e_i + \mathfrak{o} f_i) \perp H_i$$

and

$$\mathfrak{o} e_i + \mathfrak{o} f_i = \langle \delta_i \pi^{u(L_i)}, \varepsilon_i \pi^{v(L_i)} \rangle_{s(L_i)}$$

where H_i is zero or the orthogonal sum of hyperbolic planes $\pi^{s(L_i)}\mathbf{H}$ for $1 \le i \le t$.

In this part, we assume that the Jordan splitting of L satisfies one of the following conditions
- $\dim(L_i) \ge 4$ for some $1 \le i \le t$
- $u(L_i) = s(L_i)$ for some $1 \le i \le t$
- $u(L_i) \equiv u(L_j) \mod 2$ for some $i < j$ and

$$u(L_i) \ge u(L_j) \quad \text{or} \quad u(L_j) - u(L_i) \ge 2(s(L_j) - s(L_i)).$$

By [Xu, Prop. 1 and 2], Lemma 1.2.2 and Remark 1.2.3, one has

$$\mathfrak{u}\dot{F}^2 \subseteq \theta(SO(L)).$$

THEOREM 1.4.1. $\theta(SO(L)) = \dot{F}$ if one of the following conditions holds
1) $u(L_i) < v(L_i) \le s(L_i)$ for some $1 \le i \le t$;
2) $u(L_i) \equiv u(L_j) + 1 \mod 2$ for some $1 \le i < j \le t$.
Otherwise $\theta(SO(L)) = \mathfrak{u}\dot{F}^2$.

PROOF. First we show that $\theta(SO(L)) = \dot{F}$ if one of the conditions holds.
1) If $u(L_i) < v(L_i) \le s(L_i)$, then the result follows from [Xu, Prop. 2] for $\dim(L_i) \ge 4$. For $\dim(L_i) = 2$, one has

$$\theta(SO(L_i)) = G(FL_i) \not\supseteq \mathfrak{u}\dot{F}^2$$

by [X, Prop. 3] and Lemma 1.1.4. Then $\theta(SO(L)) = \dot{F}$.
2) If $u(L_i) \equiv u(L_j) + 1 \mod 2$, then

$$\delta_i \delta_j \pi^{u(L_i)+u(L_j)} \in \theta(SO(L)).$$

Therefore $\theta(SO(L)) = \dot{F}$.

Otherwise, one has

$$u(L_i) \equiv u(L_j) \mod 2 \quad \text{for } 1 \le i, j \le t$$

and

$$v(L_j) > s(L_j) \quad \text{or} \quad v(L_j) = s(L_j) = u(L_j) \quad \text{for } 1 \le j \le t.$$

If $u(L_j) + u(L_i) \le 2s(L_i)$ for some $i < j$, one has

$$(c3) \qquad d(\delta_j \delta_i) > \min\{s(L_i) - \frac{u(L_j) + u(L_i)}{2}, \ s(L_j) - u(L_j), \ s(L_i) - u(L_i)\}$$

by Definition 1.2.4.3)iv). Then we will show

$$\theta(SO(L)) \subseteq \mathfrak{u}\dot{F}^2.$$

1.4. COMPUTATIONS OF SPINOR NORMS, I

It is clear that
$$\theta(SO(L_j)) \subseteq u\dot{F}^2 \quad \text{for} \quad 1 \leq j \leq t$$
by [Xu, Prop. 2 and 3].

Let
$$\tau_\eta \in S(L) \quad \text{and} \quad \eta = \sum_{j=1}^{t} \eta_j$$
where
$$\eta_j = x_j e_j + y_j f_j + z_j g_j \in L_j$$
and g_j is a primitive vector in H_j. Then

(c4) $\quad ord(Q(\eta)) \leq \min_{1 \leq j \leq t} \{ord(x_j) + s(L_j),\ ord(y_j) + s(L_j),\ ord(z_j) + s(L_j)\}.$

Let
$$T = \{k :\ ord(Q(\eta_k)) \leq ord(Q(\eta))\ \text{for}\ 1 \leq k \leq t\}.$$
Suppose there is $k \in T$ such that
$$ord(Q(\eta_k)) = \min\{s(L_k) + ord(x_k),\ s(L_k) + ord(y_k),\ s(L_k) + ord(z_k)\}.$$
Then
$$ord(Q(\eta)) = ord(Q(\eta_k)) \quad \text{and} \quad \theta(\tau_\eta \tau_{\eta_k}) \in u\dot{F}^2$$
by (c4) and the result follows. Therefore we assume that for all $k \in T$,
$$ord(Q(\eta_k)) < \min\{s(L_k) + ord(x_k),\ s(L_k) + ord(y_k),\ s(L_k) + ord(z_k)\}.$$
This implies that

(c5) $\quad ord(Q(\eta_k)) = u(L_k) + 2ord(x_k) \quad \text{and} \quad ord(x_k) \leq s(L_k) - u(L_k).$

For any $k', k \in T$, one has
$$u(L_{k'}) + u(L_k)$$
$$\leq \min\{u(L_{k'}) + (ord(Q(\eta)) - 2ord(x_k)),\ (ord(Q(\eta)) - 2ord(x_{k'})) + u(L_k)\}$$
$$\leq \min\{u(L_{k'}) + ord(x_{k'}) + s(L_{k'}),\ u(L_k) + ord(x_k) + s(L_k)\}$$
$$\leq \min\{2s(L_{k'}), 2s(L_k)\}$$
by (c4) and (c5). Furthermore we claim that one can assume that
$$ord(Q(\eta)) < \min\{s(L_k) + ord(x_k) + ord(y_k),\ 2ord(y_k) + v(L_k),\ 2ord(z_k) + s(L_k)\}$$
for all $k \in T$.

Indeed, suppose
$$ord(Q(\eta)) = s(L_k) + ord(x_k) + ord(y_k).$$
Then
$$ord(x_k) = ord(y_k) = 0, \quad ord(Q(\eta)) = s(L_k) \quad \text{and} \quad ord(Q(\eta_k)) = u(L_k)$$
by (c4) and (c5).

If there is $1 \leq l \neq k \leq t$ such that
$$ord(Q(\eta_l)) \leq ord(Q(\eta_k)) = u(L_k),$$
then $l \in T$ and
$$ord(Q(\eta)) = s(L_k) \leq ord(x_l) + s(L_l)$$
by (c4). Therefore
$$u(L_k) \geq u(L_l) + 2ord(x_l) \geq \max\{u(L_l),\ u(L_l) + 2(s(L_k) - s(L_l))\}$$

by (c5). By Definition 1.2.4.3)ii) and iii), one has
$$ord(Q(\eta)) = s(L_k) \equiv u(L_k) = ord(Q(\eta_k)) \mod 2$$
and the result follows.

Otherwise, one has
$$ord(Q(\eta_l)) > ord(Q(\eta_k))$$
for all $1 \leq l \neq k \leq t$. Therefore
$$ord(Q(\eta)) = ord(Q(\eta_k))$$
and the result follows.

Suppose
$$ord(Q(\eta)) = 2ord(y_k) + v(L_k).$$
Then
$$s(L_k) = u(L_k) = v(L_k) \quad \text{and} \quad ord(y_k) = 0$$
by (c4). Therefore
$$ord(Q(\eta_k)) = ord(Q(\eta)) = s(L_k)$$
and the result follows.

Suppose
$$ord(Q(\eta)) = 2ord(z_k) + s(L_k).$$
Then
$$\dim(L_k) \geq 4 \quad \text{and} \quad ord(Q(\eta)) = s(L_k)$$
by (c4). By Definition 1.2.4.1), one has
$$u(L_k) \equiv s(L_k) = ord(Q(\eta)) \mod 2$$
and the claim follows.

Let i be the smallest integer such that
$$ord(Q(\eta_i)) = \min \{ord(Q(\eta_j)) : 1 \leq j \leq t\}$$
and
$$\delta_k \delta_i^{-1} = \alpha_k^2 + \beta_k^2 \pi^{d(\delta_k \delta_i)} \quad \text{with } \alpha_k, \beta_k \in \mathfrak{u}$$
for all $k \in T$ and $k \neq i$. For $k = i$, we write $\alpha_k = 1$. Then
$$2ord(x_k) + u(L_k) \geq 2ord(x_i) + u(L_i)$$
for all $k \in T$ by (c5). This implies
$$2ord(x_k) + u(L_k) + s(L_i) - \frac{u(L_k) + u(L_i)}{2} \geq ord(x_k) + ord(x_i) + s(L_i)$$
and
$$2ord(x_k) + u(L_k) + s(L_i) - u(L_i) \geq 2ord(x_i) + s(L_i).$$
Then
$$(c6) \quad 2ord(x_k) + u(L_k) + d(\delta_k \delta_i) > \min\{ord(x_k) + s(L_k), ord(x_i) + s(L_i)\}$$

by (c3). Therefore

$$\begin{aligned}
ord(Q(\eta)) &= ord(Q(\sum_{k \in T} \eta_k)) = ord(\sum_{k \in T} \delta_k \pi^{u(L_k)} x_k^2) \\
&= ord(\sum_{k \in T} \delta_k \delta_i^{-1} \pi^{u(L_k)} x_k^2) = ord(\sum_{k \in T} \alpha_k^2 \pi^{u(L_k)} x_k^2) \\
&= u(L_i) + 2 ord(x_i) + 2 ord(\sum_{k \in T} \alpha_k \pi^{(u(L_k) - u(L_i))/2} (x_k/x_i)) \\
&\equiv ord(Q(\eta_i)) \mod 2
\end{aligned}$$

by the above claim and (c6). The proof is complete. $\square$

1.5. Computations of spinor norms, II

In this section, we consider the remaining case
- $\dim(L_i) = 2$ and $u(L_i) < s(L_i)$ for $1 \leq i \leq t$.
- $0 < u(L_j) - u(L_i) < 2(s(L_j) - s(L_i))$ for $1 \leq i < j \leq t$.

The last condition holds automatically when $u(L_i) + u(L_j) \equiv 1 \mod 2$ by Definition 1.2.4.2). Write
$$L_i = \mathfrak{o}e_i + \mathfrak{o}f_i \cong (\delta_i \pi^{u(L_i)}, \varepsilon_i \pi^{v(L_i)})_{s(L_i)}$$
for $1 \leq i \leq t$. Define $(1 + \mathfrak{p}^d)\dot{F}^2 = \dot{F}$ if $d \leq 0$.

LEMMA 1.5.1. *If* $u(L_i) + u(L_j) \equiv 1 \mod 2$ *for some* $1 \leq i < j \leq t$, *then*
$$(1 + \mathfrak{p}^{u(L_j)+u(L_i)-2s(L_i)})\dot{F}^2 \subseteq \theta(SO(L)).$$

PROOF. Let
$$d = u(L_j) + u(L_i) - 2s(L_i).$$
The result follows from Lemma 1.2.1 for $d \leq 0$. We assume d is a positive odd integer. By Lemma 1.1.3, for any $\lambda \in \mathfrak{u}$, there exist $\alpha, \beta \in \mathfrak{u}$ such that
$$1 + \lambda \pi^d = \alpha^2 + \beta^2 \delta_j \delta_i^{-1} \pi^d.$$
Let
$$\eta = \alpha \pi^{s(L_i)-u(L_i)} e_i + \beta e_j.$$
Then
$$Q(\eta) = \pi^{2s(L_i)-u(L_i)} \delta_i (1 + \lambda \pi^d)$$
and
$$ord_\mathfrak{p}(\langle \eta, L \rangle) = \min\{2s(L_i) - u(L_i), \ s(L_j)\} \geq 2s(L_i) - u(L_i).$$
Therefore
$$\tau_\eta \in O(L) \quad \text{and} \quad 1 + \lambda \pi^d \in \theta(SO(L)).$$
□

LEMMA 1.5.2. *If* $u(L_i) \equiv u(L_j) \mod 2$ *and* $u(L_j) \leq s(L_i)$ *for some* $i < j$, *then*
$$(1 + \mathfrak{p}^{v(L_i)-s(L_i)})\dot{F}^2 \subseteq \theta(SO(L)).$$

PROOF. The result follows from Lemma 1.2.2 for $v(L_i) \leq s(L_i)$.
Suppose
$$v(L_i) \geq 2s(L_i) - u(L_i).$$
Then
$$v(L_i) - s(L_i) \geq s(L_i) - u(L_i) > 0$$
and the result follows from [Xu, Prop. 3].

Therefore we can assume that $v(L_i) > s(L_i)$ and $v(L_i) \equiv u(L_i) + 1 \mod 2$.

Suppose $u(L_i) \equiv s(L_i) \mod 2$. By Lemma 1.1.3, for any $\lambda \in \mathfrak{u}$ there exist $\alpha, \beta \in \mathfrak{u}$ such that
$$1 + \lambda \pi^{v(L_i)-s(L_i)} = \alpha^2 + \beta^2 \varepsilon_i \delta_j^{-1} \pi^{v(L_i)-s(L_i)}.$$
Let
$$x = \alpha \pi^{(s(L_i)-u(L_j))/2} \quad \text{and} \quad \eta = \beta f_i + x e_j.$$
Then
$$Q(\eta) = \delta_j \pi^{s(L_i)}(1 + \lambda \pi^{v(L_i)-s(L_i)}).$$

Therefore
$$\tau_\eta \in O(L) \quad \text{and} \quad (1+\mathfrak{p}^{v(L_i)-s(L_i)})\dot{F^2} \subseteq \theta(SO(L)).$$
Suppose $u(L_i) \equiv s(L_i)+1 \mod 2$. By the same argument as above we have
$$(1+\mathfrak{p}^{v(L_i)-s(L_i)+1})\dot{F^2} \subseteq \theta(SO(L)).$$
The result follows from
$$(1+\mathfrak{p}^{v(L_i)-s(L_i)+1})\dot{F^2} = (1+\mathfrak{p}^{v(L_i)-s(L_i)})\dot{F^2}.$$
$\square$

LEMMA 1.5.3. *If $u(L_i) \equiv u(L_j) \mod 2$ and $u(L_j) > s(L_i)$ for some $i < j$, then*
$$(1+\mathfrak{p}^{u(L_j)+v(L_i)-2s(L_i)})\dot{F^2} \subseteq \theta(SO(L)).$$

PROOF. Let
$$n = u(L_j) + v(L_i) - 2s(L_i).$$
If $v(L_i) \geq 2s(L_i) - u(L_i)$, then $n \geq s(L_i) - u(L_i)$ and the result follows from [Xu, Prop. 3]. Therefore we can assume that
$$v(L_i) \equiv u(L_i)+1 \mod 2.$$
For any $\lambda \in \mathfrak{u}$, there exist $\alpha, \beta \in \mathfrak{u}$ such that
$$1 + \lambda\pi^n = \alpha^2 + \beta^2 \delta_j^{-1} \varepsilon_i \pi^n.$$
Let
$$\eta = \beta\pi^{u(L_j)-s(L_i)}f_i + \alpha e_j.$$
Then
$$Q(\eta) = \delta_j \pi^{u(L_j)}(1+\lambda\pi^n).$$
Therefore
$$\tau_\eta \in O(L) \quad \text{and} \quad 1+\lambda\pi^n \in \theta(SO(L)).$$
$\square$

For any $1 \leq i < j \leq t$, we denote
$$\begin{cases} g_{ij} = v(L_i) + u(L_j) - 2s(L_i) \\ h_{ij} = v(L_j) + u(L_i) - 2s(L_i) \\ k_{ij} = u(L_i) + u(L_j) - 2s(L_i) \\ l_{ij} = d(\delta_j \delta_i) + \frac{k_{ij}}{2} \\ m_{ij} = \max\{g_{ij},\, v(L_i) - s(L_i)\} \\ n_{ij} = \max\{h_{ij},\, v(L_j) - s(L_j)\}. \end{cases}$$

LEMMA 1.5.4. *If $u(L_i) \equiv u(L_j) \mod 2$ and $u(L_i) + u(L_j) \leq 2s(L_i)$ for some $i < j$, then*
$$(1+\mathfrak{p}^d)\dot{F^2} \subseteq \theta(SO(L))$$
where
$$d = \min\{\frac{u(L_j)-u(L_i)}{2},\ s(L_j)-s(L_i)-\frac{u(L_j)-u(L_i)}{2},\ l_{ij},\ m_{ij},\ n_{ij}\}.$$

PROOF. Let
$$m = d(\delta_j \delta_i), \quad \delta_j \delta_i^{-1} = \alpha_0^2 + \beta_0^2 \pi^m \quad \text{and} \quad s = \frac{-k_{ij}}{2}$$
where $\alpha_0, \beta_0 \in \mathfrak{u}$. Since
$$0 < u(L_j) - u(L_i) < 2(s(L_j) - s(L_i)),$$
one has
$$\min\{s, \, s(L_i) - u(L_i), \, s(L_j) - u(L_j)\} = s.$$
The result follows from Lemma 1.2.2 for $m \leq s$. Therefore we assume $m > s$.

Step 1: $(1 + \mathfrak{p}^{l_{ij}})\dot{F}^2 \subseteq \theta(SO(L))$.

It is clear that we can assume that $m < \infty$. By Lemma 1.1.3, for any $\lambda \in \mathfrak{u}$, there exist $\alpha, \beta \in \mathfrak{u}$ such that
$$1 + \lambda \pi^{m-2[s/2]} = \alpha^2 + \beta^2 \pi^{m-2[s/2]}.$$
Let
$$x = (\alpha \pi^{[s/2]} + \alpha_0 \beta \beta_0^{-1}) \pi^{\frac{u(L_j) - u(L_i)}{2}} \quad \text{and} \quad y = \beta \beta_0^{-1}$$
and
$$\eta = x e_i + y e_j \in L.$$
Then
$$Q(\eta) = \delta_i \pi^{u(L_j) + 2[s/2]}(1 + \lambda \pi^{m-2[s/2]})$$
and
$$ord_{\mathfrak{p}}(\langle \eta, L \rangle) \geq \min\{\, s(L_i) + (u(L_j) - u(L_i))/2, \, s(L_j)\} \geq u(L_j) + s.$$
Therefore
$$\tau_\eta \in O(L) \quad \text{and} \quad 1 + \lambda \pi^{m-2[s/2]} \in \theta(SO(L)).$$
The result follows from
$$(1 + \mathfrak{p}^{m-s})\dot{F}^2 = (1 + \mathfrak{p}^{m-2[s/2]})\dot{F}^2.$$

Step 2: $(1 + \mathfrak{p}^{m_{ij}})\dot{F}^2 \cup (1 + \mathfrak{p}^{n_{ij}})\dot{F}^2 \subseteq \theta(SO(L))$.

It is clear that
$$m_{ij} = \begin{cases} v(L_i) - s(L_i) & \text{if } u(L_j) \leq s(L_i) \\ g_{ij} & \text{if } u(L_j) > s(L_i). \end{cases}$$
Then
$$(1 + \mathfrak{p}^{m_{ij}})\dot{F}^2 \subseteq \theta(SO(L))$$
by Lemma 1.5.2 and Lemma 1.5.3.

By considering the dual lattice, we have
$$(1 + \mathfrak{p}^{n_{ij}})\dot{F}^2 \subseteq \theta(SO(L)).$$

Step 3: $(1 + \mathfrak{p}^{\frac{u(L_j) - u(L_i)}{2}})\dot{F}^2 \cup (1 + \mathfrak{p}^{s(L_j) - s(L_i) - \frac{u(L_j) - u(L_i)}{2}})\dot{F}^2 \subseteq \theta(SO(L))$.

If $u(L_j) \leq s(L_i)$, we can assume
$$\frac{u(L_j) - u(L_i)}{2} < \begin{cases} v(L_i) - s(L_i) & \text{if } u(L_i) \equiv s(L_i) \mod 2 \\ v(L_i) - s(L_i) + 1 & \text{otherwise} \end{cases}$$
by Lemma 1.5.2. One can further assume that $u(L_j) < s(L_i)$ by [Xu, Prop. 3].

When $u(L_i) \equiv s(L_i) \mod 2$, for any $\lambda \in \mathfrak{u}$, there exists $y \in \mathfrak{u}$ such that
$$y^2 \varepsilon_i \pi^{v(L_i)-s(L_i)-\frac{u(L_j)-u(L_i)}{2}} + (\alpha_0 + \pi^{\frac{s(L_i)-u(L_j)}{2}})y$$
$$= \beta_0^2 \delta_i \pi^{m-s} + \delta_i \lambda$$
by Hensel's lemma. Let
$$x = (\alpha_0 + \pi^{\frac{s(L_i)-u(L_j)}{2}})\pi^{\frac{u(L_j)-u(L_i)}{2}}$$
and
$$\eta = xe_i + yf_i + e_j.$$
Then
$$Q(\eta) = \delta_i \pi^{s(L_i)}(1 + \lambda \pi^{\frac{u(L_j)-u(L_i)}{2}}).$$
Therefore
$$\tau_\eta \in O(L) \quad \text{and} \quad (1 + \mathfrak{p}^{\frac{u(L_j)-u(L_i)}{2}})\dot{F}^2 \subseteq \theta(SO(L)).$$
When $u(L_i) \equiv s(L_i) + 1 \mod 2$, for any $\lambda \in \mathfrak{u}$, there exists $y \in \mathfrak{u}$ such that
$$y^2 \varepsilon_i \pi^{v(L_i)-s(L_i)+1-\frac{u(L_j)-u(L_i)}{2}} + (\alpha_0 + \pi^{\frac{s(L_i)-u(L_j)+1}{2}})y$$
$$= \beta_0^2 \delta_i \pi^{m-s-1} + \delta_i \lambda$$
by Hensel's lemma. Let
$$x = (\alpha_0 + \pi^{\frac{s(L_i)-u(L_j)+1}{2}})\pi^{\frac{u(L_j)-u(L_i)}{2}}$$
and
$$\eta = xe_i + y\pi f_i + e_j.$$
Then
$$Q(\eta) = \delta_i \pi^{s(L_i)+1}(1 + \lambda \pi^{\frac{u(L_j)-u(L_i)}{2}}).$$
Therefore
$$\tau_\eta \in O(L) \quad \text{and} \quad (1 + \mathfrak{p}^{\frac{u(L_j)-u(L_i)}{2}})\dot{F}^2 \subseteq \theta(SO(L)).$$
If $u(L_j) > s(L_i)$, we can assume that
$$\min\{g_{ij}, m\} > \frac{u(L_j) - u(L_i)}{2}$$
by Lemma 1.5.3 and Step 1. By Hensel's lemma there exists $z \in \mathfrak{u}$ such that
$$z^2 \varepsilon_i \delta_i^{-1} \pi^{g_{ij}-\frac{u(L_j)-u(L_i)}{2}} + (1 + \alpha_0)\delta_i^{-1} z$$
$$= \beta_0^2 \pi^{m-\frac{u(L_j)-u(L_i)}{2}} + \lambda$$
for any $\lambda \in \mathfrak{u}$. Let
$$w = (1 + \alpha_0)\pi^{\frac{u(L_j)-u(L_i)}{2}}$$
and
$$\eta = we_i + z\pi^{u(L_j)-s(L_i)} f_i + e_j.$$
Then
$$Q(\eta) = \delta_i \pi^{u(L_j)}(1 + \lambda \pi^{\frac{u(L_j)-u(L_i)}{2}}).$$
Therefore
$$\tau_\eta \in O(L) \quad \text{and} \quad (1 + \mathfrak{p}^{\frac{u(L_j)-u(L_i)}{2}})\dot{F}^2 \subseteq \theta(SO(L)).$$
By considering the dual lattice, one has
$$(1 + \mathfrak{p}^{s(L_j)-s(L_i)-\frac{u(L_j)-u(L_i)}{2}})\dot{F}^2 \subseteq \theta(SO(L)).$$

The proof is complete. □

LEMMA 1.5.5. *If $u(L_i) \equiv u(L_j) \mod 2$ and $u(L_i) + u(L_j) > 2s(L_i)$ for some $i < j$, then*
$$1 + \mathfrak{p}^d \subseteq \theta(SO(L))$$
where
$$d = \min\{k_{ij} + d(\delta_i \delta_j),\ g_{ij},\ h_{ij}\}.$$

PROOF. Let
$$n = k_{ij} + d(\delta_j \delta_i).$$
For any $\lambda \in \mathfrak{u}$, there exist $\alpha,\ \beta \in \mathfrak{u}$ such that
$$1 + \lambda \pi^n = \alpha^2 + \beta^2 \delta_j \delta_i^{-1} \pi^{k_{ij}}.$$
by (c1) and Lemma 1.1.3. Let
$$\eta = \alpha \pi^{s(L_i) - u(L_i)} e_i + \beta e_j.$$
Then
$$Q(\eta) = \pi^{2s(L_i) - u(L_i)} \delta_i (1 + \lambda \pi^n)$$
and
$$ord_{\mathfrak{p}}(\langle \eta, L \rangle) = \min\{2s(L_i) - u(L_i), s(L_j)\} \geq 2s(L_i) - u(L_i).$$
Therefore
$$\tau_\eta \in O(L) \quad \text{and} \quad 1 + \mathfrak{p}^n \subseteq \theta(SO(L)).$$
If $v(L_j) > s(L_j)$, then
$$h_{ij} > s(L_j) - u(L_j).$$
The result follows from [Xu, Prop. 3] and Lemma 1.1.4.

Otherwise, $v(L_j) \leq s(L_j)$. So we have
$$v(L_j) \equiv u(L_j) + 1 \mod 2.$$
By Lemma 1.1.3, for any $\lambda \in \mathfrak{u}$ there exist $\alpha, \beta \in \mathfrak{u}$ such that
$$1 + \lambda \pi^{h_{ij}} = \alpha^2 + \beta^2 \varepsilon_j \delta_i^{-1} \pi^{h_{ij}}.$$
Let
$$x = \alpha \pi^{s(L_i) - u(L_i)} \quad \text{and} \quad \eta = x e_i + \beta f_j.$$
Then
$$ord(Q(\eta)) = 2s(L_i) - u(L_i)$$
and
$$ord_{\mathfrak{p}}(\langle \eta, L \rangle) = \min\{2s(L_i) - u(L_i), s(L_j)\} \geq 2s(L_i) - u(L_i).$$
Therefore
$$\tau_\eta \in O(L) \quad \text{and} \quad (1 + \mathfrak{p}^{h_{ij}}) \subseteq \theta(SO(L)).$$
By considering the dual lattice, one has
$$(1 + \mathfrak{p}^{g_{ij}}) \subseteq \theta(SO(L))$$
and the result follows. □

1.5. COMPUTATIONS OF SPINOR NORMS, II

REMARK 1.5.6. 1) If
$$d(\delta_j\delta_i) < \min\{v(L_j) - u(L_i),\ 2(s(L_j) - s(L_i)) + v(L_i) - u(L_j)\},$$
then $d(\delta_i\delta_j)$ is independent of the choice of the norm generators of L_i and L_j.

2) For any two-dimensional lattice K with $u(K) < s(K)$, we have
$$\theta(SO(K)) \supseteq \begin{cases} (1+\mathfrak{p}^{s(K)-u(K)})\dot{F}^2 & \text{if } v(K) > s(K) \\ (1+\mathfrak{p}^{-Ord(\Delta(FK))+1})\dot{F}^2 & \text{otherwise} \end{cases}$$
by Lemma 1.1.4 and [Xu, Prop. 3].

3) If $u(L_i) \equiv u(L_j) \mod 2$ and $u(L_i) + u(L_j) > 2s(L_i)$, then
$$(1+\mathfrak{p}^{\frac{u(L_j)-u(L_i)}{2}})\dot{F}^2 \cup (1+\mathfrak{p}^{s(L_j)-s(L_i)-\frac{u(L_j)-u(L_i)}{2}})\dot{F}^2 \subseteq \theta(SO(L)).$$

We first show that
$$(1+\mathfrak{p}^{\frac{u(L_j)-u(L_i)}{2}})\dot{F}^2 \subseteq \theta(SO(L)).$$

Since
$$\frac{u(L_j) - u(L_i)}{2} > s(L_i) - u(L_i),$$
we only need to consider the case that $v(L_i) \leq s(L_i)$ by the above 2). Since
$$\frac{u(L_j) - u(L_i)}{2} + \frac{u(L_j) - u(L_i)}{2} = g_{ij} - Ord(\Delta(FL_i)),$$
we have
$$\frac{u(L_j) - u(L_i)}{2} \geq g_{ij}$$
or
$$\frac{u(L_j) - u(L_i)}{2} > -Ord(\Delta(FL_i)).$$
The result follows from Lemma 1.5.5 and the above 2).

By considering the dual lattice, one has
$$(1+\mathfrak{p}^{s(L_j)-s(L_i)-\frac{u(L_j)-u(L_i)}{2}})\dot{F}^2 \subseteq \theta(SO(L)).$$
$\square$

Define
- $d_1 = \min_{i<j}\{k_{ij} : u(L_j) \equiv u(L_i) + 1 \mod 2\}$.
- $d_2 = \min_{i<j}\{d_{ij} : u(L_j) \equiv u(L_i) \mod 2 \text{ and } u(L_i) + u(L_j) \leq 2s(L_i)\}$
  with
$$d_{ij} = \min\{\frac{u(L_j)-u(L_i)}{2},\ s(L_j) - s(L_i) - \frac{u(L_j)-u(L_i)}{2},\ l_{ij},\ m_{ij},\ n_{ij}\}.$$
- $d_3 = \min_{i<j}\{d_{ij} : u(L_j) \equiv u(L_i) \mod 2 \text{ and } u(L_i) + u(L_j) > 2s(L_i)\}$
  with
$$d_{ij} = \min\{\frac{u(L_j)-u(L_i)}{2},\ s(L_j) - s(L_i) - \frac{u(L_j)-u(L_i)}{2},\ k_{ij} + d(\delta_i\delta_j),\ g_{ij},\ h_{ij}\}.$$

Let
$$G = \prod_{i=1}^{3}(1+\mathfrak{p}^{d_i})$$
and
$$N = \{Q(e_{i_1})\cdots Q(e_{i_{2r}}) : 1 \leq i_1 \leq i_2 \leq \cdots \leq i_{2r} \leq t\}.$$

THEOREM 1.5.7. $\theta(SO(L)) = GN \prod_{i=1}^{t} \theta(SO(L_i))$.

PROOF. It is clear that we only need to show that
$$\theta(SO(L)) \subseteq GN \prod_{i=1}^{t} \theta(SO(L_i))$$
by lemmas and Remark 1.5.6 in this section.

Suppose $u(L_i) + u(L_j) \leq 2s(L_i)$ for some $i < j$ and one of the following conditions holds
$$\begin{cases} u(L_i) \equiv u(L_j) + 1 \mod 2 \\ d(\delta_i \delta_j) \leq s(L_i) - \frac{u(L_i)+u(L_j)}{2} \\ v(L_i) \leq s(L_i) \text{ or } v(L_j) \leq s(L_j). \end{cases}$$
Then the result follows from Lemma 1.5.1 and Lemma 1.5.4. Therefore, we assume that the Jordan splitting of L does not satisfy the above conditions. This implies that
$$d_1 > 0, \quad d_2 > 0, \quad \text{and} \quad d_3 > 0.$$

Let $\tau_\eta \in S(L)$ and
$$\eta = \Sigma_{j=1}^{t} \eta_j \quad \text{where} \quad \eta_j = x_j e_j + y_j f_j \in L_j.$$
Then
(c7) $$\phi = ord(Q(\eta)) \leq \min_{1 \leq j \leq t}\{ord(x_j) + s(L_j),\ ord(y_j) + s(L_j)\}.$$

Let i be the smallest integer such that
$$ord(Q(\eta_i)) = \min_{1 \leq j \leq t}\{ord(Q(\eta_j))\}.$$
Then $ord(Q(\eta_i)) \leq ord(Q(\eta))$ and
(c8) $$ord(Q(\eta_i)) \leq \min_{1 \leq j \leq t}\{ord(x_j) + s(L_j),\ ord(y_j) + s(L_j)\}$$
and $\tau_{\eta_i} \in O(L_i)$.

By the domination principle in [Ri1], one has
(c9) $$ord(Q(\eta_i)) \leq 2ord(x_j) + u(L_j)$$
for any $1 \leq j \leq t$ when $v(L_j) < \infty$. For $v(L_j) = \infty$, the above inequality (c9) still holds by (c8).

Case I: $v(L_i) > s(L_i)$.

First we claim that
$$ord(Q(\eta_i)) = 2ord(x_i) + u(L_i) \quad \text{and} \quad ord(x_i) \leq s(L_i) - u(L_i).$$

For $v(L_i) < \infty$, the claim follows from (c8) and the domination principle in [Ri1].

Otherwise one only needs to consider
$$v(L_i) = \infty \quad \text{and} \quad ord(Q(\eta_i)) \geq ord(x_i) + ord(y_i) + s(L_i).$$
Then $ord(x_i) = ord(y_i) = 0$ by (c8) and the claim follows as well.

Secondly we show
$$ord(Q(\eta_j)) - \phi \geq d_1 \quad \text{for all } j \text{ with } u(L_j) + u(L_i) \equiv 1 \mod 2.$$
Indeed, for $j < i$, one gets
$$\min\{2ord(x_j) + u(L_j), 2ord(y_j) + v(L_j), ord(x_j) + ord(y_j) + s(L_j)\} - \phi$$
$$\geq \min\{\phi + u(L_j) - 2s(L_j), \phi + v(L_j) - 2s(L_j), ord(x_j)\}$$
$$\geq \min\{u(L_i) + u(L_j) - 2s(L_j), ord(Q(\eta_i)) - s(L_j)\} \geq k_{ij}$$
by (c7), (c8) and the above claim.

For $j > i$, one gets
$$\phi \leq ord(x_i) + s(L_i) \leq 2s(L_i) - u(L_i)$$
by (c7) and the above claim. Then
$$ord(Q(\eta_j)) - \phi \geq k_{ij}$$
and the result follows.

Finally, let
$$S = \{1 \leq l \leq t : u(L_l) \equiv u(L_i) \mod 2\}$$
and
$$\delta_j \delta_i^{-1} = \alpha_j^2 + \beta_j^2 \pi^{d(\delta_i \delta_j)} \quad \text{with } \alpha_j, \beta_j \in \mathfrak{u}$$
where $j \in S$ and $j \neq i$. Assume $\alpha_i = 1$ and $\beta_i = 0$. Then
$$\sum_{l \in S} Q(\eta_l) = Q(\gamma_i) + \delta_i \sum_{l \in S} x_l^2 \beta_l^2 \pi^{u(L_l) + d(\delta_i \delta_l)}$$
$$+ \sum_{l \in S, l \neq i} \alpha_l y_i x_l \pi^{s(L_i) + (u(L_l) - u(L_i))/2}$$
$$+ \sum_{l \in S, l \neq i} x_l y_l \pi^{s(L_l)} + \sum_{l \in S, l \neq i} y_l^2 \varepsilon_l \pi^{v(L_l)}$$
where
$$\gamma_i = (\sum_{l \in S} x_l \alpha_l \pi^{\frac{u(L_l) - u(L_i)}{2}}) e_i + y_i f_i \in L_i$$
by (c9) and the above claim.

Let $l \in S$ and $l \neq i$.

1) Estimate $A = 2ord(x_l) + u(L_l) + d(\delta_i \delta_l) - \phi$.
When $u(L_l) + u(L_i) > 2\min\{s(L_l), s(L_i)\}$, then
$$A \geq \begin{cases} \phi + u(L_l) - 2s(L_l) + d(\delta_i \delta_l) & \text{if } l < i \\ u(L_l) + d(\delta_i \delta_l) - ord(x_i) - s(L_i) & \text{if } l > i \end{cases}$$
$$\geq u(L_l) + u(L_i) + d(\delta_i \delta_l) - 2\min\{s(L_l), s(L_i)\}$$
$$\geq d_3$$
by (c7) and the above claim.

When $u(L_l) + u(L_i) \leq 2\min\{s(L_l),\ s(L_i)\}$, then

$$A \geq \begin{cases} ord(x_l) + u(L_l) - s(L_l) + d(\delta_i\delta_l) & \text{if } l < i \\ ord(x_l) + u(L_l) + d(\delta_i\delta_l) - s(L_i) - ord(x_i) & \text{if } l > i \end{cases}$$

$$\geq d(\delta_i\delta_l) + \frac{u(L_i) + u(L_l)}{2} - \min\{s(L_l),\ s(L_i)\}$$

$$\geq d_2$$

by (c7), (c9) and the above claim.

2) Estimate $B = ord(y_i) + ord(x_l) + (u(L_l) - u(L_i))/2 + s(L_i) - \phi$.

$$B \geq \begin{cases} s(L_i) - s(L_l) - \frac{u(L_i) - u(L_l)}{2} & \text{if } l < i \\ \frac{u(L_l) - u(L_i)}{2} & \text{if } l > i \end{cases}$$

$$\geq \min\{d_2,\ d_3\}$$

by (c7).

3) Estimate $C = ord(x_l) + ord(y_l) + s(L_l) - \phi$.
If $l < i$, then

$$C \geq ord(x_l) \geq (u(L_i) - u(L_l))/2 \geq \min\{d_2,\ d_3\}$$

by (c7), (c9) and the above claim.
If $l > i$, then

$$C \geq ord(x_l) + s(L_l) - ord(x_i) - s(L_i)$$

$$\geq s(L_l) - s(L_i) - \frac{u(L_l) - u(L_i)}{2}$$

$$\geq \min\{d_2,\ d_3\}$$

by (c7), (c9) and the above claim.

4) Estimate $D = 2ord(y_l) + v(L_l) - \phi$.
If $l < i$, then

$$D \geq \max\{\phi + v(L_l) - 2s(L_l),\ v(L_l) - s(L_l)\}$$

$$\geq \max\{g_{li},\ v(L_l) - s(L_l)\}$$

$$\geq \min\{m_{li},\ d_3\} \geq \min\{d_2,\ d_3\}$$

by (c7) and the above claim.
If $l > i$, then

$$D \geq \max\{v(L_l) - ord(x_i) - s(L_i),\ v(L_l) - s(L_l)\}$$

$$\geq \max\{h_{il},\ v(L_l) - s(L_l)\}$$

$$\geq \min\{n_{il},\ d_3\} \geq \min\{d_2,\ d_3\}$$

by (c7) and the above claim.

Since

$$\sum_{l \in S} Q(\eta_l) = Q(\eta) + \sum_{l \notin S} Q(\eta_l),$$

we have

$$ord(Q(\gamma_i)) = \phi \quad \text{and} \quad Q(\eta)^{-1}Q(\gamma_i) \in G$$

1.5. COMPUTATIONS OF SPINOR NORMS, II

by the above arguments. For any $l \in S$, one has
$$ord(x_l \alpha_l \pi^{\frac{u(L_l) - u(L_i)}{2}}) + s(L_i) \geq ord(x_i) + s(L_i) \geq \phi$$
by (c9), the above claim and (c7). Therefore
$$\tau_{\gamma_i} \in O(L_i)$$
and the result follows.

Case II: $v(L_i) \leq s(L_i)$.

We can assume
$$ord(Q(\eta_i)) = 2ord(y_i) + v(L_i)$$
by the domination principle in [Ri1] and the same argument as that in the case I. Then
$$ord(y_i) \leq s(L_i) - v(L_i)$$
by (c8).

If $u(L_j) + u(L_i) \equiv 1 \mod 2$ and $j < i$, one gets
$$\min\{2ord(x_j) + u(L_j),\ 2ord(y_j) + v(L_j),\ ord(x_j) + ord(y_j) + s(L_j)\} - \phi$$
$$\geq \min\{\phi + u(L_j) - 2s(L_j),\ \phi + v(L_j) - 2s(L_j),\ ord(x_j)\}$$
$$\geq \min\{k_{ji},\ ord(Q(\eta_i)) - s(L_j)\} = k_{ji} \geq d_1$$
by (c7) and (c8).

If $u(L_j) + u(L_i) \equiv 1 \mod 2$ and $j > i$, one has
$$ord(Q(\eta_j)) - \phi \geq u(L_j) - s(L_i) - ord(y_i)$$
$$\geq u(L_j) + v(L_i) - 2s(L_i) \geq k_{ij} \geq d_1$$
by (c7).

If $u(L_j) + u(L_i) \equiv 0 \mod 2$, then
$$u(L_i) + u(L_j) > 2\min\{s(L_i),\ s(L_j)\}$$
by our assumption.

When $j < i$, then
$$\min\{2ord(x_j) + u(L_j),\ 2ord(y_j) + v(L_j),\ ord(x_j) + ord(y_j) + s(L_j)\} - \phi$$
$$\geq \min\{\phi + u(L_j) - 2s(L_j),\ \phi + v(L_j) - 2s(L_j),\ ord(x_j)\}$$
$$\geq \min\{v(L_i) + u(L_j) - 2s(L_j),\ v(L_i) + v(L_j) - 2s(L_j),\ \frac{v(L_i) - u(L_j)}{2}\}$$
$$\geq d_3$$
by (c7) and (c9).

When $j > i$, then
$$ord(Q(\eta_j)) - \phi \geq u(L_j) - s(L_i) - ord(y_i)$$
$$\geq u(L_j) + v(L_i) - 2s(L_i) \geq k_{ij} \geq d_3$$
by (c7).

By the above argument, we have
$$ord(Q(\eta_i)) = \phi \quad \text{and} \quad Q(\eta_i)Q(\eta)^{-1} \in G.$$

The proof is complete. $\square$

REMARK 1.5.8. Suppose $s(L) = 0$ and $\dim(L) = 2t$ with $t \geq 2$. If $ord(\mathfrak{v}(L)) < 3t(t-1)$, then $\mathfrak{u}\dot{F}^2 \subseteq \theta(SO(L))$.

PROOF. It is clear that we can assume that L is not of W-type and the Jordan decomposition
$$L = L_1 \perp \cdots \perp L_t$$
satisfies the assumption of this section.

If $s(L) = 0$, $\dim(L) = 4$ and $ord(\mathfrak{v}(L)) \leq 4$, then
$$s(L_1) = 0, \quad s(L_2) = 2 \quad \text{and} \quad u(L_2) - u(L_1) = 2.$$
Therefore
$$\mathfrak{u}\dot{F}^2 \subseteq \theta(SO(L))$$
by Lemma 1.5.4 and Remark 1.5.6. 3).

Since
$$ord(\mathfrak{v}(L)) = 2(s(L_2) + \cdots + s(L_t)) < 3t(t-1),$$
there exists some $1 \leq i < t$ such that
$$2(s(L_{i+1}) - s(L_i)) \leq 4.$$
Then
$$\theta(SO(L)) \supseteq \theta(SO(L_i \perp L_{i+1})) \supseteq \mathfrak{u}\dot{F}^2$$
and the result follows. □

We point out that the bound $ord(\mathfrak{v}(L)) < 3t(t-1)$ given in the above remark cannot be unconditionally improved in view of the following example.

EXAMPLE 1.5.9. Suppose $L = L_1 \perp \cdots \perp L_t$ where $L_i = (\pi^{-4(t-i)}, 0)_{3(i-1)}$. Then we have $\theta(SO(L)) = (1 + \mathfrak{p}^2)\dot{F}^2$.

PROOF. It is obvious that L is not of W-type. The result follows from Theorem 1.5.7 and [Xu, Prop. 3]. □

1.6. Group structure of $\theta(X(L/K))$

Suppose L and K are two lattices in V where $V = FL$ and $K \subseteq L$. It is clear that
$$X(L/K)SO(L) = SO(K)X(L/K) = X(L/K).$$

THEOREM 1.6.1. $\theta(X(L/K))$ is a group.

PROOF. Since
$$SO(L) \subseteq X(L/K) \quad \text{and} \quad SO(K) \subseteq X(L/K),$$
one only needs to consider
$$[\dot{F} : \theta(SO(L))] > 2 \quad \text{and} \quad [\dot{F} : \theta(SO(K))] > 2.$$
So we can assume that neither L nor K is of W-type. Let
$$L = L_1 \perp L_2 \perp \cdots \perp L_t \quad \text{and} \quad K = K_1 \perp K_2 \perp \cdots \perp K_s$$
be the Jordan splittings of L and K respectively and
$$L_1^\perp = L_2 \perp \cdots \perp L_t \quad \text{and} \quad K_1^\perp = K_2 \perp \cdots \perp K_s.$$
Then they are in the case of §1.5. Write
$$L_1 = \mathfrak{o}e_1 + \mathfrak{o}f_1 \quad \text{and} \quad K_1 = \mathfrak{o}g_1 + \mathfrak{o}h_1$$
where e_1 and g_1 are the norm generators of L_1 and K_1 respectively.

Suppose $u(L) < u(K)$. Then there is
$$L \supset L' = (\mathfrak{p}e_1 + \mathfrak{o}f_1) \perp L_1^\perp \supseteq K$$
such that $X(L'/K) = X(L/K)$. By repeating the above argument, we can assume
$$u(L) = u(K) \quad \text{and} \quad e_1 = g_1.$$
Suppose $s(L) < s(K)$. Let
$$L \supset L' = (\mathfrak{o}e_1 + \mathfrak{p}f_1) \perp L_1^\perp \supseteq K.$$
For any $\sigma \in X(L/K)$, there is $\tau \in O(L)$ such that $\tau\sigma^{-1}(e_1) = e_1$ by Corollary 1.3.2. Write
$$\tau\sigma^{-1}(y) = ae_1 + bf_1 + z$$
for any $y \in K$, where $a, b \in \mathfrak{o}$, $z \in L_1^\perp$. Then
$$b = (\langle e_1, f_1 \rangle)^{-1} \langle e_1, y \rangle \in \mathfrak{p} \quad \text{and} \quad \sigma\tau^{-1} \in X(L'/K).$$
Therefore
$$X(L/K) = X(L'/K)SO(L).$$
By repeating this argument, we can assume that
$$s(L) = s(K) \quad \text{and} \quad L_1 = K_1.$$
By Proposition 1.3.1, one has
$$X(L/K) = X(L_1^\perp/K_1^\perp)SO(L).$$
The result follows from induction. □

1.7. Reduction formula for $\theta(X(L/K))$, I

Suppose
$$L = L_1 \perp \cdots \perp L_t$$
is a Jordan splitting of a lattice L. For an integer $1 \leq i \leq t$,
$$L(i) = \{x \in L : \langle x, L \rangle \subseteq \pi^{s(L_i)}\mathfrak{o}\}$$
is called the $i - th$ invariant sublattice of L. It is clear that
$$L = L(1) \supseteq \cdots \supseteq L(t)$$
and
$$u(L) = u(L(1)) \leq \cdots \leq u(L(t)).$$

We define a set of integers inductively
$$1 \leq j_L(1) < j_L(2) < \cdots < j_L(l) \leq t$$
as follows
$$j_L(1) = \min\{1 \leq i \leq t : u(L(i)) < u(L(i+1))\}$$
and $j_L(a)$ is defined as the minimal integer i with $j_L(a-1) < i \leq t$ satisfying
$$u(L(i)) < u(L(i+1))$$
and
$$u(L(i)) - u(L(i-1)) < 2(s(L_i) - s(L_{i-1}))$$
for $a > 1$. It is clear that
$$\{j_L(1), \cdots, j_L(l)\} \quad \text{and} \quad \{u(L_{j_L(1)}), \cdots, u(L_{j_L(l)})\}$$
are independent of the Jordan splitting of L.

DEFINITION 1.7.1. The set of integers defined as above
$$\{j_L(1), \cdots, j_L(l)\}$$
is called the jumping set of L and is denoted by $J(L)$. The corresponding the scales and norms
$$\{s(L_{j_L(1)}), \cdots, s(L_{j_L(l)})\}$$
and
$$\{u(L_{j_L(1)}), \cdots, u(L_{j_L(l)})\}$$
are called the jumping scales and norms of L respectively.

We denote
$$j_L(i) = +\infty, \quad L_{j_L(i)} = 0 \quad \text{and} \quad u(L_{j_L(i)}) = +\infty$$
if $i > l$.

It is clear that
$$u(L_{j_L(1)}) = u(L) < u(L_j)$$
for $j > j_L(1)$. Since
$$u(L(i)) = \min\{2(s(L_i) - s(L_{i-1})) + u(L(i-1)), u(L_i), \cdots, u(L_t)\}$$
for $i > 1$, we have
$$u(L_{j_L(a)}) = u(L(j_L(a))).$$

Furthermore

(c10) $\quad j \in J(L) \Leftrightarrow \begin{cases} u(L_j) < u(L_i) + 2(s(L_j) - s(L_i)) & \text{for all } 1 \leq i < j \\ u(L_j) < u(L_k) & \text{for all } j < k \leq t. \end{cases}$

For any $1 \leq m \leq t$, we define

$$j_L(m^-) = \max\{j \in J(L) : j \leq m\} \quad \text{and} \quad j_L(m^+) = \min\{j \in J(L) : j \geq m\}.$$

LEMMA 1.7.2. *Let $K \subseteq L' \subseteq L$ with $\dim(L) = \dim(L')$. Suppose L' is split by $\pi^{r'}\mathbf{H}$ and L is split by $\pi^r\mathbf{H}$.*
If $r' \equiv r + 1 \mod 2$, then $\theta(X(L/K)) = \dot{F}$.

PROOF. It is clear that

$$\theta(X(L/K)) \supseteq \mathfrak{u}\dot{F}^2.$$

Write

$$\pi^r\mathbf{H} = \mathfrak{o}e + \mathfrak{o}f \quad \text{with} \quad Q(e) = Q(f) = 0 \quad \text{and} \quad \langle e, f \rangle = \pi^r$$

and

$$\pi^{r'}\mathbf{H} = \mathfrak{o}e' + \mathfrak{o}f' \quad \text{with} \quad Q(e') = Q(f') = 0 \quad \text{and} \quad \langle e', f' \rangle = \pi^{r'}.$$

The result follows from

$$\tau_{e'+f'}\tau_{e+f} \in X(L/K) \quad \text{and} \quad \theta(\tau_{e'+f'}\tau_{e+f}) = \pi^{r+r'}.$$

$\square$

Before we establish the reduction formula of computing $\theta(X(L/K))$ for two lattices $K \subseteq L$ with $s(L) < s(K)$, we need the following lemma.

LEMMA 1.7.3. *Suppose $s(K) > s(L)$. Then $\theta(X(L/K)) = \dot{F}$ if one of the following conditions holds*
  1): $j_L(1) > 1$;
  2): $\dim(L_1) \geq 4$;
  3): $u(K) > u(L)$ and L_1 can be split by a hyperbolic plane.

PROOF. We can assume that L is not of W-type. Then there is a Jordan splitting of L with the first Jordan component L_1 such that

$$L_1 = (\mathfrak{o}e_1 + \mathfrak{o}f_1) \perp M \quad \text{with} \quad Q(e_1) = Q(f_1) = 0 \quad \text{and} \quad \langle e_1, f_1 \rangle = 1$$

by scaling. Let $L_1^\perp$ is the orthogonal complement of L_1 in L.
  If

$$K \subseteq (\mathfrak{p}e_1 + \mathfrak{o}f_1) \perp M \perp L_1^\perp,$$

the result follows from Lemma 1.7.2.
  Otherwise, there exists $u \in K$ such that

$$u = ae_1 + bf_1 + z$$

where $z \in M \perp L_1^\perp$, $a \in \mathfrak{u}$ and $b \in \mathfrak{o}$. There exists a lattice N such that

$$L = (\mathfrak{o}u + \mathfrak{o}f_1) \perp N.$$

Since $s(K) > 0$, one has

$$K \subseteq (\mathfrak{o}u + \mathfrak{p}f_1) \perp N = L'.$$

1)If $j_L(1) > 1$, then one has that L' is of W-type or split by $\pi\mathbf{H}$ by Remark 1.2.3. The result follows from Lemma 1.7.2.
  2)If $dim(L_1) \geq 4$, then let N_1 be the first Jordan component of N.

If $u(N_1) \geq ord(Q(u))$, then $j_{L'}(1) > 1$. Replacing L by L', one obtains the result by induction.

Otherwise, one has that
$$N_1 \perp (\mathfrak{o}u + \mathfrak{p}f_1)$$
is of W-type or split by $\pi\mathbf{H}$. The result follows from Lemma 1.7.2.

3) If $u(K) > u(L)$ and L_1 can be split by a hyperbolic plane, one only needs to consider that $j_L(1) = 1$ and L_1 is a hyperbolic plane. Since
$$u(K) > u(L) = s(L),$$
we have that
$$\mathfrak{o}u + \mathfrak{p}f_1 \cong \pi\mathbf{H}$$
and the result follows from Lemma 1.7.2. □

LEMMA 1.7.4. *Suppose $u(L) < s(L)$ or L_1 can not be split by any hyperbolic plane. Then the set*
$$\{x \in L : \quad Q(x) \in \mathfrak{pn}(L)\}$$
is a sublattice of L.

PROOF. When $u(L) < s(L)$, there are sublattices
$$M = \mathfrak{o}x + \mathfrak{o}y$$
with
$$ord(Q(x)) = u(L) \quad \text{and} \quad ord(Q(y)) > u(L)$$
and N with $u(N) > u(L)$ of L such that
$$L = M \perp N.$$
Then
$$\{x \in L : \quad Q(x) \in \mathfrak{pn}(L)\} = (\mathfrak{p}x + \mathfrak{o}y) \perp N$$
is a sublattice of L.

Otherwise one can write
$$L_1 = \mathfrak{o}u + \mathfrak{o}v \cong (1,\rho)_{s(L)}$$
and
$$u(L_2 \perp \cdots \perp L_t) > u(L).$$
Therefore
$$\{x \in L : \quad Q(x) \in \mathfrak{pn}(L)\} = (\mathfrak{p}u + \mathfrak{p}v) \perp L_2 \perp \cdots \perp L_t$$
is a sublattice of L. The proof is complete. □

RF I. *Suppose $s(L) < s(K)$ and neither of them is of W-type.*

If $\dim(L_1) = 2$ and $j_L(1) = 1$ and

1) When $u(L) < u(K)$ and L_1 is not a hyperbolic plane, then
$$X(L/K) = X(L'/K)$$
where
$$K \subseteq L' = \{x \in L : \quad Q(x) \in \mathfrak{pn}(L)\} \subset L.$$
2) When $u(L) = u(K)$ and $s(L_2) \geq s(K)$, then
$$X(L/K) = X(L'/K)SO(L)$$
where
$$K \subseteq L' = \{x \in L : \quad \langle x, K \rangle \subseteq \mathfrak{s}(K)\} \subset L.$$

Otherwise $\theta(X(L/K)) = \dot{F}$.

PROOF. It is clear that we can assume that $s(L) = 0$.
First we show 1) and 2). Write $L_1 = \mathfrak{o}e_1 + \mathfrak{o}f_1$ such that
$$ord(Q(e_1)) = u(L_1), \quad ord(Q(f_1)) = v(L_1) \quad \text{and} \quad \langle e_i, f_1 \rangle = 1.$$
Let $L_1^\perp$ be the orthogonal complement of L_1 in L. Since $j_L(1) = 1$, one has
$$u(L_1) < u(L_1^\perp).$$
1) When $u(L) < u(K)$ and L_1 is not a hyperbolic plane, then
$$L' = \{x \in L : Q(x) \in \mathfrak{pn}(L)\}$$
is a sublattice of L by Lemma 1.7.4 and
$$X(L/K) = X(L'/K)$$
by the standard order consideration.

2) When $u(L) = u(K)$ and $s(L_2) \geq s(K)$, one can assume that $e_1 \in K$. For any $\eta \in K$, there are $x, y \in \mathfrak{o}$ and $z \in L_1^\perp$ such that
$$\eta = xe_1 + yf_1 + z.$$
Since
$$y = \langle \eta, e_1 \rangle \in \mathfrak{s}(K),$$
one has
$$K \subseteq L' = (\mathfrak{o}e_1 + \mathfrak{s}(K)f_1) \perp L_1^\perp \subset L.$$
For any $\sigma \in X(L/K)$, one has $\sigma^{-1}e_1 \in L$. Then there is $\tau \in SO(L)$ such that
$$\tau\sigma^{-1}e_1 = e_1$$
by Corollary 1.3.2 since $\dim(L_1) = 2$ and $j_L(1) = 1$. For any $\eta \in K$, there are $a, b \in \mathfrak{o}$ and $w \in L_1^\perp$ such that
$$\tau\sigma^{-1}\eta = ae_1 + bf_1 + w.$$
Then
$$b = b\langle e_1, f_1 \rangle = \langle \tau\sigma^{-1}\eta, e_1 \rangle = \langle \eta, e_1 \rangle \in \mathfrak{s}(K).$$
Therefore
$$\sigma\tau^{-1} \in X(L'/K).$$
So
$$X(L/K) = X(L'/K)SO(L).$$
Conversely, by Lemma 1.7.3, one only needs to show that
$$\theta(X(L/K) = \dot{F}$$
if
$$u(K) = u(L), \quad s(L_2) < s(K) \quad \text{and} \quad \dim(L_1) = 2.$$
By the same argument as that in 2), there is a lattice M with
$$K \subset M \subset L \quad \text{and} \quad s(M) = s(L_2)$$
such that the rank of the first Jordan component of $M \geq 4$. Then
$$\theta(X(L/K)) \supseteq \theta(X(M/K)) = \dot{F}$$
by Lemma 1.7.3. The proof is complete. □

Finally we determine the jumping set of the new lattice obtained by RF I which is quite useful in the following sections for computing the relative spinor norms.

REMARK 1.7.5. In RF I, one has
$$j_L(2) > 2 \quad \text{or} \quad s(L_2) > s(L_1) + 1.$$

For case 1):
If $u(L_{j_L(2)}) = u(L_1) + 2$, then
$$J(L') = \{j_L(2) - \delta, \cdots, j_L(l) - \delta\}.$$
If $u(L_{j_L(2)}) > u(L_1) + 2$, then
$$J(L') = \{1, j_L(2) - \delta, \cdots, j_L(l) - \delta\}.$$

Here
$$\delta = \begin{cases} 0 & \text{when } s(L_2) > s(L_1) + 1 \\ 1 & \text{when } s(L_2) = s(L_1) + 1. \end{cases}$$

For case 2):

$$J(L') = \{1\} \cup \{j - \delta : j \in J(L), j > 1 \text{ with } u(L_j) < u(L_1) + 2(s(L_j) - s(K))\}$$
where
$$\delta = \begin{cases} 0 & \text{when } s(L_2) > s(K) \\ 1 & \text{when } s(L_2) = s(K). \end{cases}$$

PROOF. Suppose
$$j_L(2) = 2 \quad \text{and} \quad s(L_2) = s(L_1) + 1.$$
Then
$$u(L_1) < u(L_2) < u(L_1) + 2.$$
This implies that $u(L_2) = u(L_1) + 1$ and L is of W-type.

For case 1), it is clear that
$$u(L_{j_L(2)}) \geq u(L_1) + 2$$
by Def.1.2.4. 2) since L is not of W-type.

Let $i > 1$ and $i \notin J(L)$. Then
$$\begin{cases} i \notin J(L') & \text{if } s(L_2) > s(L_1) + 1 \\ i - 1 \notin J(L') & \text{if } s(L_2) = s(L_1) + 1 \end{cases}$$
by (c10).

For any $j \in J(L)$ and $j > 1$, one has
$$u(L_j) < u(L_1) + 2(s(L_j) - s(L_1))$$
by (c10). Suppose
$$u(L_j) = u(L_1) + 2(s(L_j) - s(L_1)) - 1.$$
Then
$$u(L_1) + u(L_j) = 2u(L_j) - 2s(L_j) + 2s(L_1) + 1 \leq 2s(L_1) + 1$$
and L is of W-type. Therefore
(c11) $$u(L_j) \leq u(L_1) + 2(s(L_j) - s(L_1)) - 2.$$

Since
$$u(L') = u(L_1) + 1 \quad \text{or} \quad u(L') = u(L_1) + 2,$$
the result follows from (c10) and (c11).

For case 2), it is clear that
$$\begin{cases} i \notin J(L') & \text{if } s(K) < s(L_2) \\ i-1 \notin J(L') & \text{if } s(K) = s(L_2) \end{cases}$$
for $1 < i \leq t$ and $i \notin J(L)$ by (c10). Then the result follows from (c10). $\square$

1.8. Reduction formula for $\theta(X(L/K))$, II

Suppose K and L are two lattices with $K \subseteq L$. We need to establish a reduction formula for computing $\theta(X(L/K))$ in the case of $s(L) = s(K)$.

LEMMA 1.8.1. *Suppose $\pi^r \mathbf{H}$ splits $\sigma(\pi^r \mathbf{H} \perp L)$ for some*

$$\sigma \in X(\pi^r \mathbf{H} \perp L / \pi^r \mathbf{H} \perp K).$$

Then there is $\tau \in SO(\pi^r \mathbf{H} \perp L)$ such that

$$\sigma\tau|_{\pi^r \mathbf{H}} = 1_{\pi^r \mathbf{H}} \quad \text{and} \quad \sigma\tau|_L \in X(L/K).$$

PROOF. It follows from the cancellation theorem for hyperbolic planes (see [Po]). $\square$

Let

$$K = K_1 \perp K_2 \perp \cdots \perp K_s$$

and

$$L = L_1 \perp L_2 \perp \cdots \perp L_t$$

be the Jordan splitting of K and L. If $K \subseteq L$ and $s(L) = s(K)$, one can assume that $K_1 \subseteq L_1$. In fact K_1 splits L_1 in this case. We denote $K_1^\perp$ and $L_1^\perp$ as the orthogonal complements of K_1 and L_1 in K and L respectively.

RF II. *Suppose $s(L) = s(K) = 0$ and neither of them is of W-type.*

1) If K is split by a unimodular hyperbolic plane, then there are two lattices $K' \subseteq L'$ such that

$$L = \mathbf{H} \perp L' \quad \text{and} \quad K = \mathbf{H} \perp K'$$

and

$$X(L/K) = X(L'/K')SO(L).$$

2) Suppose K is not split by any unimodular hyperbolic plane and $\dim(L_1) = \dim(K_1)$.

i) If $j_L(1) > 1$ or one of the following conditions holds

$$\begin{cases} u(L_{j_L(2)}) - u(L) \le 2s(L_2) \\ \dim(L_2) = 2 \quad \text{and} \quad u(L_2) = u(K_1^\perp) = u(L) + 2s(L_2) \end{cases}$$

where $L_2 \neq 0$, then

$$X(L/K) = X(L_1^\perp / K_1^\perp)SO(L).$$

ii) If $j_L(1) = 1$, $u(L_{j_L(2)}) - u(L) > 2s(L_2)$ and $s(L_2) = s(K_2)$, then

$$X(L/K) = X(L'/K')SO(L)$$

where

$$L = \pi^{s(L_2)} \mathbf{H} \perp L' \quad \text{and} \quad K = \pi^{s(K_2)} \mathbf{H} \perp K'.$$

3) Suppose K is not split by any unimodular hyperbolic plane, $j_L(1) = 1$ and $\dim(L_1) = \dim(K_1) + 2$.

i) If $u(L) < u(K)$, then

$$X(L/K) = X(L'/K)$$

where

$$K \subseteq L' = \{x \in L : Q(x) \in \mathfrak{pn}(L)\} \subset L.$$

ii) If $u(L) = u(K) = s(K)$ and $Ord(\Delta(FL_1)) = \infty$, then

$$X(L/K) = X(L'/K')SO(L)$$

where
$$L = K_1 \perp L' \quad \text{and} \quad K = K_1 \perp K'.$$

Otherwise $\theta(X(L/K)) = \dot{F}$.

PROOF. By Lemma 1.8.1, we only need to prove 2) and 3).

For convenience, we rewrite Proposition 1.3.1 as follows.

PROPOSITION (1.3.1'). *Suppose L is not of W-type and $\dim(L_1) = 2$. Assume that one of the following conditions holds*
$$\begin{cases} j_L(1) > 1 \\ u(L) + s(L_2) \geq 2s(L_1) \\ u(L_{j_L(2)}) - u(L) \leq 2(s(L_2) - s(L_1)). \end{cases}$$
If $\sigma L_1 \subseteq L$ for some $\sigma \in O(FL)$, then there is $\tau \in E(L)$ such that $\tau\sigma|_{L_1} = 1$.

For 2), it is clear that i) follows from Proposition 1.3.1' and Proposition 1.3.4 by choosing a vector in $K_1^\perp$ as a norm generator of L_2 when $u(K_1^\perp) = u(L_2)$.

For ii), one has
$$u(K_1^\perp) \geq u(L_1^\perp) \quad \text{and} \quad u(K) = u(K_1) = u(L_1) = u(L)$$
and
$$u(K_1) \equiv u(K_1^\perp) \mod 2$$
by Definition 1.2.4.2). It is clear that
$$(c12) \qquad u(L_i) \geq \min\{u(L_1) + 2s(L_i),\ u(L_{j_L(2)})\}$$
for $1 < i \leq t$. Therefore
$$u(K_1^\perp) - u(K) \geq u(L_1^\perp) - u(L_1) = \min_{1 < i \leq t}\{u(L_i)\} - u(L_1) \geq 2s(L_2) = 2s(K_2)$$
by (c12). This implies K is split by
$$\pi^{s(K_1^\perp)}\mathbf{H}.$$
Since $\dim(K_1) = \dim(L_1)$, we have
$$\pi^{s(K_1^\perp)}\mathbf{H}$$
splits L as well. Therefore ii) follows from Lemma 1.8.1.

For 3), it is clear that i) follows from the norm consideration and Lemma 1.7.4.
For ii), one can write
$$K_1 = \mathfrak{o}e_1 + \mathfrak{o}f_1 = (1, \rho)_0 \quad \text{and} \quad M = \mathfrak{o}e + \mathfrak{o}f = (1, \rho)_0$$
where $L_1 = K_1 \perp M$. For any $\sigma^{-1} \in X(L/K)$, one has
$$\sigma e_1 = x_1 e_1 + y_1 f_1 + xe + yf + z \in L$$
and at least one of x_1, y_1, x and y is a unit.
Claim: There is $\varrho \in SO(L)$ such that $\varrho\sigma e_1 = e_1$.

First we can assume that $y_1 \in \mathfrak{p}$. Otherwise one can take
$$\varrho = \tau_{e_1}\tau_{\sigma e_1 + e_1} \in SO(L).$$
We can further assume that $x_1 \in \mathfrak{p}$. Otherwise
$$\tau_{e_1+f_1} \in O(L) \quad \text{and} \quad \tau_{e_1+f_1}\sigma e_1 = x_1' e_1 + y_1' f_1 + z' \in L$$

with $y'_1 \in \mathfrak{u}$ and reduce to the above argument. Then
$$x \in \mathfrak{u} \quad \text{or} \quad y \in \mathfrak{u}.$$
Since $E^{e_1+e}_{f_1+f} \in O(L)$ and
$$E^{e_1+e}_{f_1+f} e = e_1 \quad \text{and} \quad E^{e_1+e}_{f_1+f} f = f_1,$$
the claim follows from replacing σ with $E^{e_1+e}_{f_1+f}\sigma$.

By the above claim, we can assume that $\sigma e_1 = e_1$. Then there are α, β and γ in $\mathfrak{o}$ and $w \in L_1^\perp$ such that
$$\sigma f_1 = \alpha e_1 + f_1 + \beta e + \gamma f + w \in L.$$
If $\alpha \in \mathfrak{u}$, then
$$\tau_{\sigma f_1 + f_1} \in O(L) \quad \text{and} \quad \tau_{\sigma f_1 + f_1}\sigma|_{K_1} = 1.$$
Otherwise the result follows from replacing σ with $\tau_{e_1+\pi e}\sigma$ where $\tau_{e_1+\pi e} \in O(L)$.

Conversely, we will show that $\theta(X(L/K)) = \dot{F}$ if the conditions in RF II are not satisfied. Let $L_1 = M \perp K_1$.

- $\dim(M) \geq 4$; or $\dim(M) = 2$ and $u(M) \geq u(L_1^\perp)$.
  By Lemma 1.7.3, one has
  $$\theta(X(M \perp L_1^\perp / K_1^\perp)) = \dot{F}.$$
  The result follows from
  $$X(M \perp L_1^\perp / K_1^\perp) \subseteq X(L/K).$$

- $\dim(M) = 2$ and $u(L_1^\perp) > u(M)$.
  It is clear that $j_L(1) = 1$ and L_1 is split by $\mathbf{H}$ by Definition 1.2.4.1). By 3), one has
  $$u(M) \geq u(K_1)$$
  and
  $$Ord(\Delta(FL_1)) < \infty \quad \text{when } u(K) = s(K).$$
  Write
  $$M = \mathfrak{o}g + \mathfrak{o}h \quad \text{with} \quad ord(Q(g)) = u(M) \quad \text{and} \quad ord(Q(h)) = v(M).$$
  If $u(M) < u(K_1^\perp)$, then
  $$K_1^\perp \subseteq (\mathfrak{p}g + \mathfrak{o}h) \perp L_1^\perp$$
  and
  $$Ord(\Delta(FM)) = \infty \quad \text{when } u(K) = s(K).$$
  Therefore
  $$K_1 \perp (\mathfrak{p}g + \mathfrak{o}h)$$
  is of W-type or split by $\pi\mathbf{H}$ by Remark 1.2.3. The result follows from Lemma 1.7.2.
  If $u(M) \geq u(K_1^\perp)$, one has
  $$u(K_1^\perp) \geq u(M \perp L_1^\perp) = u(M).$$

Then $u(M) = u(K_1^\perp)$. Therefore we can assume that $g \in K_1^\perp$. It is clear that
$$K_1^\perp \subseteq (\mathfrak{o}g + \mathfrak{p}h) \perp L_1^\perp.$$
Since
$$X(K_1 \perp (\mathfrak{o}g + \mathfrak{p}h) \perp L_1^\perp/K) \subseteq X(L/K),$$
and
$$K_1 \perp (\mathfrak{o}g + \mathfrak{p}h)$$
is of W-type or split by $\pi\mathbf{H}$ by Remark 1.2.3, the result follows from Lemma 1.7.2.

- $M = 0$.

  By 2), we have
  $$j_L(1) = 1 \quad \text{and} \quad u(L_{j_L(2)}) > u(L_1) + 2s(L_2)$$
  and
  $$s(L_2) < s(K_2)$$
  and one of the following conditions is satisfied
  $$\begin{cases} \dim(L_2) \geq 4; \text{ or} \\ u(L_2) \neq u(K_1^\perp); \text{ or} \\ u(L_2) - u(L_1) > 2s(L_2). \end{cases}$$
  It is clear that $j_L(2) > 2$ by (c10) and
  $$\pi^{s(L_2)}\mathbf{H}$$
  splits L by Definition 1.2.4.1) and 2) and
  $$\min\{u(L_2), u(K_1^\perp)\} \geq u(L_1^\perp).$$
  By Lemma 1.7.3, we can assume that $L_2 = \mathfrak{o}e_2 + \mathfrak{o}f_2$ where
  $$ord(Q(e_2)) = u(L_2) < u(L_3 \perp \cdots \perp L_t)$$
  and
  $$v(L_2) = ord(Q(f_2)).$$
  Then
  $$u(K_1^\perp) \geq u(L_1^\perp) = u(L_2).$$
  When $u(L_2) \neq u(K_1^\perp)$, we have
  $$u(L_2) < u(K_1^\perp)$$
  and
  $$K \subseteq L' = L_1 \perp (\mathfrak{p}e_2 + \mathfrak{o}f_2) \perp L_3 \perp \cdots \perp L_t \subseteq L$$
  by norm consideration. Since
  $$u(L_2) - u(L_1) \geq 2s(L_2),$$
  we get
  $$L_1 \perp (\mathfrak{p}e_2 + \mathfrak{o}f_2)$$
  is of W-type or split by
  $$\pi^{s(L_2)+1}\mathbf{H}$$
  by Remark 1.2.3. The result follows from Lemma 1.7.2. Therefore we can further assume that $u(L_2) = u(K_1^\perp)$ and $e_2 \in K_1^\perp$.

Since $s(K_2) < s(L_2)$, one gets
$$K \subseteq L' = L_1 \perp (\mathfrak{o}e_2 + \mathfrak{p}f_2) \perp L_3 \perp \cdots \perp L_t \subseteq L.$$
When $u(L_2) - u(L_1) > 2s(L_2)$, then
$$L_1 \perp (\mathfrak{o}e_2 + \mathfrak{p}f_2)$$
is of W-type or split by
$$\pi^{s(L_2)+1}\mathbf{H}$$
by Remark 1.2.3. The result follows from Lemma 1.7.2. □

For computing the relative spinor norms, we determine the jumping set for the new lattice obtained by RF II.

REMARK 1.8.2. For case 1) of RF II, one has
$$J(L') = \begin{cases} J(L) & \text{if } \dim(L_1) > 2 \\ \{j_L(\delta) - 1, \cdots, j_L(l) - 1\} & \text{otherwise} \end{cases}$$
where
$$\delta = \begin{cases} 1 & \text{if } j_L(1) > 1 \\ 2 & \text{otherwise.} \end{cases}$$
For case i) in 2) of RF II, one has
If $j_L(1) > 1$, then
$$J(L_1^\perp) = \{j_L(1) - 1, j_L(2) - 1, \cdots j_L(l) - 1\}.$$
If $j_L(1) = 1$ and $u(L_{j_L(2)}) \leq u(L_1) + 2s(L_2)$, then
$$J(L_1^\perp) = \{j_L(2) - 1, \cdots j_L(l) - 1\}.$$
If $j_L(1) = 1$ and $u(L_{j_L(2)}) > u(L_1) + 2s(L_2)$, then
$$J(L_1^\perp) = \{1, j_L(2) - 1, \cdots j_L(l) - 1\}.$$
For case ii) in 2) of RF II, one has $j_L(1) = 1$ and
$$J(L') = \{1, j_L(2) - \delta, \cdots j_L(l) - \delta\}$$
where
$$\delta = \begin{cases} 1 & \text{if } \dim(L_2) = 2 \\ 0 & \text{if } \dim(L_2) > 2. \end{cases}$$
For case i) in 3) of RF II, one has
$$J(L') = \begin{cases} \{2, j_L(2) + \delta, \cdots, j_L(l) + \delta\} & \text{if } u(L_{j_L(2)}) > u(L) + 2 \\ \{j_L(2) + \delta, \cdots, j_L(l) + \delta\} & \text{if } u(L_{j_L(2)}) = u(L) + 2 \end{cases}$$
where
$$\delta = \begin{cases} 0 & \text{if } s(L_2) = 1 \\ 1 & \text{if } s(L_2) > 1. \end{cases}$$
For case ii) in 3) of RF II, one has $J(L') = J(L)$.

1.8. REDUCTION FORMULA FOR $\theta(X(L/K))$, II

PROOF. For case 1), the result follows from (c10).

For i) in Case 2), the result follows from (c10) if $j_L(1) > 1$.
If
$$j_L(1) = 1 \text{ and } u(L_{j_L(2)}) \leq u(L_1) + 2s(L_2),$$
then
$$u(L_1^\perp) = u(L_{j_L(2)})$$
by (c12). Therefore
$$j_{L_1^\perp}(1) = j_L(2) - 1$$
and the result follows from (c10).
If
$$j_L(1) = 1 \text{ and } u(L_{j_L(2)}) > u(L_1) + 2s(L_2),$$
then
$$u(L_2) = u(L_1) + 2s(L_2)$$
by RF II and $j_L(2) > 2$ by (c10). Therefore
$$j_{L_1^\perp}(1) = 1$$
and the result follows from (c10).

For ii) in case 2), one has $j_L(2) > 2$ by (c10) and the result follows from (c10).

For case i) in 3), one can write
$$L_1 = K_1 \perp M \text{ and } M = \mathfrak{o}g + \mathfrak{o}h$$
where
$$u(M) = ord(Q(g)) = u(L) \text{ and } v(M) = ord(Q(h)).$$
Then
$$L' = K_1 \perp (\mathfrak{p}g + \mathfrak{o}h) \perp L_1^\perp.$$
Since L is not of W-type, one has
$$u(L') = u(L) + 2 \leq \min\{u(K_1), u(L_1^\perp)\}.$$
If $u(L') < u(L_{j_L(2)})$, then
$$j_{L'}(1) = 2 \text{ and } j_{L'}(i) = j_L(i) + \delta$$
for $2 \leq i \leq l$ by (c10).
If $u(L') = u(L_{j_L(2)})$, then
$$j_{L'}(i) = j_L(i+1) + \delta$$
for $1 \leq i \leq l - 1$ by (c10).
For case ii) in 3), the result is obvious. □

If one removes a Jordan component which is not in the jumping set or the orthogonal sublattice of such component, the jumping set of the new lattice relatively does not change.

REMARK 1.8.3. Suppose M is $\mathfrak{s}(L_i)$-modular and splits L. Let
$$L = M \perp L'.$$
If $i \notin J(L)$, then
$$J(L') = \{j_L(1), \cdots, j_L(i^-), j_L(i^+) - \delta, \cdots, j_L(l) - \delta\}$$
where
$$\delta = \begin{cases} 0 & \text{if } rank(M) < rank(L_i) \\ 1 & \text{if } rank(M) = rank(L_i). \end{cases}$$

PROOF. It is clear that
$$J(L') \supseteq \{j_L(1), \cdots, j_L(i^-), j_L(i^+) - \delta, \cdots, j_L(l) - \delta\}$$
by (c10). Since
(c13) $\quad u(L(a)) = \min\{u(L_{j_L(a^-)}) + 2(s(L_a) - s(L_{j_L(a^-)})), u(L_{j_L(a^+)})\}$
for $1 \leq a \leq t$ by (c10), we have
(c14) $\quad u(L'(a)) = \begin{cases} u(L(a)) & \text{if } a < i \\ u(L(a+\delta)) & \text{if } a \geq i. \end{cases}$

So we only need to consider that $\delta = 1$ by (c14). By Definition 1.7.1, one has $h \notin J(L)$ if and only if
$$u(L(h+1)) = u(L(h))$$
or
$$u(L(h)) = u(L(h-1)) + 2(s(L_h) - s(L_{h-1})).$$
Therefore
$$\begin{cases} h \notin J(L') & \text{if } h < i-1 \text{ and } h \notin J(L) \\ h-1 \notin J(L') & \text{if } h > i+1 \text{ and } h \notin J(L) \end{cases}$$
by (c14).

When $h = i + 1 \notin J(L)$, we only need to consider that
$$u(L(i+1)) = u(L(i)) + 2(s(L_{i+1}) - s(L_i))$$
by (c14). Since $i \notin J(L)$, one has
$$u(L(i)) = u(L(i-1)) + 2(s(L_i) - s(L_{i-1})).$$
Therefore
$$u(L'(i)) = u(L'(i-1)) + 2(s(L'_i) - s(L'_{i-1}))$$
by (c13). So
$$i \notin J(L')$$
by Definition 1.7.1.

When $h = i - 1 \notin J(L)$, one can show that
$$i - 1 \notin J(L')$$
by the same argument as above. $\square$

Finally we give an application of the reduction formulae for relative spinor norms. Let E/F be a finite extension and $\mathfrak{o}_E$ be the ring of integers of E. For a $\mathfrak{o}$-lattice L, we can lift it as a $\mathfrak{o}_E$-lattice and denote
$$\widetilde{L} = L \otimes_{\mathfrak{o}} \mathfrak{o}_E.$$
Let
$$N_{E/F} = \text{Norm from } E \text{ to } F.$$
We have the following norm principle.

PROPOSITION 1.8.4.
$$N_{E/F}(\theta(X(\widetilde{L}/\widetilde{K}))) \subseteq \theta(X(L/K)).$$

PROOF. It is clear that we can assume that neither L nor K is of W-type. Neither is $\widetilde{L}$ nor $\widetilde{K}$ by Definition 1.2.4. By Theorem 1.4.1, Theorem 1.5.7 and [Xu], we have
$$N_{E/F}(\theta(SO(\widetilde{L}))) \subseteq \theta(SO(L))$$
for any $\mathfrak{o}$-lattice L. The result follows from RF I and II. □

1.9. Some computation via reduction formulae

Suppose
$$L = L_1 \perp \cdots \perp L_t$$
is a Jordan splitting of a lattice L. For $r \in \mathbb{Z}$, we denote
$$L^r = L_1 \perp \cdots \perp L_i \quad \text{with} \quad s(L_i) \leq r < s(L_{i+1}).$$
Let $K \subseteq L$ be a sublattice of L and
$$K = K_1 \perp \cdots \perp K_s$$
be a Jordan splitting of K.

LEMMA 1.9.1. *If* $\dim(L^r) - \dim(K^r) \geq 4$ *for some* $r \in \mathbb{Z}$, *then*
$$\theta(X(L/K)) = \dot{F}.$$

PROOF. The result follows from Lemma 1.7.3 if
$$s(L_1) \leq r < s(L_2).$$
In general, by applying RF I and II, there are the sub-lattices $L' \subseteq L$ and $K' \subseteq K$ satisfying
$$\dim((L')^r) - \dim((K')^r) \geq 4$$
such that
$$X(L'/K') \subseteq X(L/K).$$
The result follows from induction on rank. □

LEMMA 1.9.2. *If there is* $1 \leq i \leq s$ *such that* $s(K_i) \geq s(L_{i+1})$, *then*
$$\theta(X(L/K)) \supseteq \mathrm{u}\dot{F}^2.$$

PROOF. When $i = 1$, then $\theta(X(L/K)) = \dot{F}$ or there is a sub-lattice $K \subseteq L' \subseteq L$ with $s(L') = s(L_2)$ such that
$$X(L'/K) \subseteq X(L/K)$$
by RF I. Then the first Jordan component of L' is of rank 4 and
$$\mathrm{u}\dot{F}^2 \subseteq \theta(X(L'/K))$$
by [Xu].

When $i > 1$, we can assume that $\dim(L_l) = \dim(K_l) = 2$ for $l < i$ by [Xu]. By RF I and II, we have
$$\theta(X(L/K)) \supseteq \mathrm{u}\dot{F}^2$$
or there exists a sublattice $L' \subseteq L$ with the same type of Jordan splitting of
$$L_i \perp \cdots \perp L_t$$
satisfying
$$L' \supseteq K_i \perp \cdots \perp K_s$$
such that
$$X(L'/K_i \perp \cdots \perp K_s) \subseteq X(L/K).$$
This reduce to the case for $i = 1$. □

1.9. SOME COMPUTATION VIA REDUCTION FORMULAE

For a lattice L, let
$$P(L) = \{x \in L : ord(Q(x)) \equiv u(L) + 1 \mod 2\}.$$

DEFINITION 1.9.3. The weight $w(L)$ of L is defined as
$$w(L) = \begin{cases} \infty & \text{if } P(L) \text{ is empty} \\ \min\{ord(Q(x)) : x \in P(L)\} & \text{otherwise} \end{cases}.$$

Since a quadratic space is universal if the dimension of the space is more than two, one has $P(L)$ is empty if and only if
$$L \cong (\delta\pi^{u(L)}, \rho\delta^{-1}\pi^{2s(L)-u(L)})_{s(L)}.$$

When $\dim(L) = 2$ and $Ord(\Delta(FL)) < 0$, one has $w(L) = v(L)$.

For any $K \subseteq L$, we denote
$$\begin{cases} u(K, L) = u(L) + 2(s(K) - s(L)) \\ \kappa(K, L) = \min\{\frac{u(K)-u(L)}{2}, s(K) - s(L)\} \end{cases}.$$

LEMMA 1.9.4. *If $w(L) \leq s(L) + \kappa(K, L)$ and*
$$\theta(X(L/K)) \supseteq u\dot{F}^2,$$
then $\theta(X(L/K)) = \dot{F}$.

PROOF. It is clear that the first Jordan component L_1 of L can be chosen as
$$u(L_1) = u(L) \quad \text{and} \quad v(L_1) = w(L).$$
Suppose $w(L) \leq s(L)$. Then
$$\theta(X(L/K)) \supseteq \theta(SO(L_1)) = G(FL_1)$$
and the result follows from Lemma 1.1.4. Therefore we can assume that $w(L) > s(L)$. This implies that $s(K) > s(L)$ by the given condition. By Lemma 1.7.3, we can further assume that $\dim(L_1) = 2$ and $j_L(1) = 1$. By applying RF I and the given condition, there is a lattice $K \subseteq L' \subset L$ such that
$$w(L') = w(L) \leq s(L) + \kappa(K, L) = s(L').$$
The result follows from the same argument as above. □

LEMMA 1.9.5. *Suppose $s(L) = s(K)$ and $\dim(L_1) = \dim(K_1)$. If $j_L(1) = 1$, then*
$$j_K(1) = 1.$$

PROOF. Suppose $j_K(1) > 1$. Then $u(K) = u(K_1^\perp)$. Since K_1 splits L as the first Jordan component of L and $j_L(1) = 1$, one has
$$L = K_1 \perp N, \quad u(N) > u(L) \quad \text{and} \quad K_1^\perp \subseteq N.$$
Then
$$u(L) = u(K_1) \geq u(K) = u(K_1^\perp) \geq u(N) > u(L).$$
A contradiction is derived. □

LEMMA 1.9.6. *Let* $j_L(1) = 1$ *and* $\dim(L^{s(K)}) = \dim(K_1)$. *Suppose*
$$\theta(X(L/K)) \neq \dot{F}.$$
Then
$$u(L_{j_L(2)}) \geq u(K,L) \quad or \quad u(L_{j_L(2)}) > u(K).$$
Moreover, if
$$u(K,L) < u(L_{j_L(2)}) \quad or \quad u(K) < u(L_{j_L(2)}),$$
then
$$u(K) = \min\{\ u(K),\ u(K,L)\ \} < u(L_{j_L(2)}).$$
If $u(K,L) = u(L_{j_L(2)}) \leq u(K)$, *then*
$$u(L) \equiv s(K) \mod 2 \quad and \quad w(L) > s(K).$$

PROOF. It is clear that neither L nor K is of W-type. Suppose
$$u(L_{j_L(2)}) < u(K,L) \quad \text{and} \quad u(L_{j_L(2)}) \leq u(K).$$
Then
$$s(K) > s(L) \quad \text{and} \quad u(L_i) \geq u(L_{j_L(2)}) \text{ for all } i > 1$$
by (c12). One can apply RF I and obtain a sequence of lattices
$$K \subseteq M_\alpha \subset \cdots \subset M_1 \subset M_0 = L$$
satisfying
$$X(M_h/K) = X(L/K) \quad \text{and} \quad s(M_h) = s(M_{h-1}) + 1$$
with $0 \leq h \leq \alpha$ and $s(M_\alpha) = s(K)$.

Case I: *There are k and k' such that $u(M_k) \equiv u(M_{k'}) + 1 \mod 2$.*

It is clear that $\pi \in \theta(X(L/K))$ and we can assume that
$$j_K(1) = j_{M_h}(1) = 1 \quad \text{for } 0 \leq h \leq \alpha.$$
Then
$$u(M_\alpha) = u(K_1) = u(K) \geq u(L_{j_L(2)}).$$
Let k is the smallest integer such that $u(M_k) \geq u(L_{j_L(2)})$. Then
$$u(M_{k-1}) + u(L_{j_L(2)}) \leq 2u(M_{k-1}) + 2 \leq 2s(M_{k-1}) + 2.$$
Since M_{k-1} is not of W-type, one has
$$u(M_{k-1}) \equiv u(L_{j_L(2)}) \mod 2 \quad \text{and} \quad u(M_k) = u(L_{j_L(2)}).$$
This implies that $j_{M_k}(1) > 1$ by Remark 1.7.5 and a contradiction is derived.

Case II: *All $u(M_h)$'s have the same parity for $0 \leq h \leq \alpha$.*

Then there is $0 \leq k \leq \alpha$ satisfying $s(M_k) < s(K)$ such that
$$u(M_k) = u(L_{j_L(2)}) \quad \text{or} \quad u(M_k) = u(L_{j_L(2)}) - 1$$
If $u(M_k) = u(L_{j_L(2)})$, a contradiction is derived from Remark 1.7.5 and Lemma 1.7.3.

If $u(M_k) = u(L_{j_L(2)}) - 1$, one has
$$u(M_k) + u(L_{j_L(2)}) = 2u(M_k) + 1 \leq 2s(M_k) + 1$$
and M_k is of W-type. A contradiction is derived.

1.9. SOME COMPUTATION VIA REDUCTION FORMULAE

Therefore
$$u(L_{j_L(2)}) \geq u(K,L) \quad \text{or} \quad u(L_{j_L(2)}) > u(K).$$

By RF I, there is a lattice $K \subseteq L' \subset L$ such that
$$X(L/K) = X(L'/K), \quad u(L') \leq u(K,L) \quad \text{and} \quad s(L') = s(K).$$
If $u(K) < u(L_{j_L(2)})$ or $u(K,L) < u(L_{j_L(2)})$, we claim that
$$u(K) \leq u(K,L).$$
Suppose not, we have
$$u(K,L) < u(L_{j_L(2)}).$$
Then $j_{L'}(1) = 1$ by Remark 1.7.5. It is obvious that the rank of the first Jordan component of L' is equal to
$$\dim(L^{s(K)}) = \dim(K_1).$$
Therefore
$$u(K) = u(K_1) = u(L') \leq u(K,L)$$
by Lemma 1.9.5. A contradiction is derived.

If $u(K,L) = u(L_{j_L(2)}) \leq u(K)$, we claim that
$$u(L') = u(K,L) = u(L_{j_L(2)}).$$
Suppose not, we have $j_{L'}(1) = 1$ by Remark 1.7.5. Since the rank of the first Jordan component of L' is equal to
$$\dim(L^{s(K)}) = \dim(K_1),$$
one has
$$u(K) = u(K_1) = u(L') < u(K,L)$$
by Lemma 1.9.5. A contradiction is derived.

By Remark 1.7.5 and the above claim, one has $j_{L'}(1) > 1$. Therefore
$$\mathfrak{u}\dot{F}^2 \subseteq \theta(X(L/K)).$$
So, by Lemma 1.9.4, $w(L) > s(K)$. Since L' is not of W-type, one has
$$u(K,L) = u(L') \equiv s(L') = s(K) \mod 2$$
by Definition 1.2.4. 1) and 3) ii). $\square$

LEMMA 1.9.7. *Let $j_L(1) = 1$ and $\dim(L^{s(K)}) = \dim(K_1) + 2$. Suppose*
$$\theta(X(L/K)) \neq \dot{F}.$$
Then
$$w(L) > s(L) + \kappa(K,L)$$
and
$$u(K,L) = \min\{u(K,L),\ u(K)\} < u(L_{j_L(2)}).$$
Furthermore, if
$$K \text{ is not split by } \pi^{s(K)}\mathbf{H} \text{ and } u(K) = u(K,L),$$
then
$$u(K) = s(K) \quad \text{and} \quad Ord(\Delta(FL^{s(K)})) = \infty.$$

PROOF. It is clear that neither L nor K is of W-type. By Lemma 1.9.1, one has
$$s(K) = s(L) \quad \text{or} \quad s(K) = s(L_2).$$
When $s(K) = s(L)$, the result follows from Lemma 1.9.4 and RF II.
When $s(K) = s(L_2)$, by Lemma 1.7.3, one has
$$\dim(L_1) = 2 \quad \text{and} \quad \dim(L_2) = \dim(K_1).$$
By Lemma 1.9.2 and Lemma 1.9.4, one also has
$$w(L) > s(L) + \kappa(K, L).$$
Suppose
$$u(L_{j_L(2)}) \leq \min\{u(K, L), u(K)\} \quad \text{or} \quad u(K) < u(K, L).$$
If
$$u(L_{j_L(2)}) \leq \min\{u(K, L), u(K)\},$$
there exists a sublattice $K \subseteq L' \subseteq L$ such that
$$X(L'/K) = X(L/K), \quad s(L') = s(K) \quad \text{and} \quad u(L') = u(L_{j_L(2)})$$
by RF I. Then $j_{L'}(1) > 1$ by Remark 1.7.5. Let
$$L' = K_1 \perp M.$$
Then $j_M(1) > 1$ by (c10). By Lemma 1.7.3,
$$\theta(X(L/K)) \supseteq \theta(X(M/K_1^\perp)) = \dot{F}.$$
A contradiction is derived.

If $u(K) < u(K, L)$, there is a sublattice $K \subseteq N \subseteq L$ such that
$$X(N/K) = X(L/K), \quad s(N) < s(K) \quad \text{and} \quad u(N) = u(K)$$
by RF I. Therefore
$$s(K) > s(N) \geq u(N) = u(K).$$
By RF II, one has
$$\theta(X(L/K)) = \theta(X(N/K)) = \dot{F}.$$
A contradiction is derived.

If K is not split by
$$\pi^{s(K)}\mathbf{H} \quad \text{and} \quad u(K) = u(K, L),$$
there is a sublattice $K \subseteq P \subseteq L$ such that
$$X(P/K) = X(L/K), \quad s(P) = s(K) \quad \text{and} \quad u(K) = u(P)$$
by RF I. The result follows from RF II. □

1.10. Some notation

In order to obtain the closed forms of $\theta(X(L/K))$ with $K \subseteq L$, we will introduce some notation in this section. Let
$$L = L_1 \perp \cdots \perp L_t$$
and
$$K = K_1 \perp \cdots \perp K_s$$
be the Jordan splittings of L and K respectively.

For our purpose of computing the relative spinor norms, we can assume that K is not of W-type. Therefore we always take the Jordan splitting of K such that

K_i is the orthogonal sum of hyperbolic planes if $i \notin J(K)$

by (c10) and Definition 1.2.4. In fact one can also establish the minimal norm Jordan splitting in the sense of [Xu2] for characteristic 2 and take such Jordan splitting for K.

For any $1 \leq b \leq s$, we denote
$$K[b] = K_b \perp \cdots \perp K_s.$$
Then
$$u(K[b]) = \min\{s(K_b),\ u(K_{j_K(b^+)})\}$$
which is independent of the choice of Jordan splitting of K.

DEFINITION 1.10.1. The lattice L is called reducible with respect to K if the sequence of lattices
$$L = M_0 \supset M_1 \supset \cdots \supset M_l \supseteq K$$
with
$$\mathfrak{n}(M_l) = \mathfrak{p}^{\kappa(K,L)}\mathfrak{n}(L)$$
which are defined inductively as the following
$$M_{i+1} = \{x \in M_i :\ Q(x) \in \mathfrak{pn}(M_i)\}$$
satisfy that

- If $\kappa(K, L) > 0$, then M_i is not of W-type whose first Jordan component is of rank two and is not hyperbolic plane and
$$\kappa(K, M_i) > 0 \quad \text{and} \quad j_{M_i}(1) = 1$$
for $0 \leq i \leq l - 1$;
- If $s(M_l) < s(K)$, then
$$j_{M_l}(1) = 1 \quad \text{and} \quad \dim(M_l^{s(K)-1}) = 2.$$

Denote
$$mid_K(L) = M_l$$
and
$$red_K(L) = \{x \in M_l :\ \langle x, K \rangle \subseteq \mathfrak{s}(K)\}.$$

Suppose L is reducible with respect to K. Then K_1 splits $red_K(L)$ and there is a sublattice $L_K\{1\}$ of L such that
$$red_K(L) = K_1 \perp L_K\{1\}.$$
Inductively, suppose $L_K\{i\}$ is reducible with respect to $K[i+1]$ for $1 \leq i \leq s-1$. Then there is a sublattice
$$L_K\{i+1\}$$
of L such that
$$red_{K[i+1]}(L_K\{i\}) = K_{i+1} \perp L_K\{i+1\}.$$

DEFINITION 1.10.2. The lattice L is called reducible with respect to K completely if K is not of W-type and $L_K\{i\}$ is reducible with respect to $K[i+1]$ for all $0 \leq i \leq s-1$ where
$$L_K\{0\} = L.$$

By Prop.1.2.7 and RF I,
$$\theta(X(L/K)) = \dot{F}$$
if L is not reducible with respect to K completely.

DEFINITION 1.10.3. A lattice N which is not of W-type is called even type if there is a Jordan decomposition
$$N = N_1 \perp \cdots \perp N_k$$
such that
$$u(N) \equiv u(N_1) \equiv \cdots \equiv u(N_k) \mod 2.$$

It should be pointed out that Def.1.10.3 is intrinsic and independent of the choice of Jordan splitting of N. Indeed, for any $i \notin J(N)$, one has
$$u(N_i) \geq \min\{u(N_{j_N(i-)}) + 2(s(N_i) - s(N_{j_N(i-)})),\ u(N_{j_N(i+)})\}$$
by (c13). By Def.1.2.4. 2), one has
$$u(N_i) \equiv \begin{cases} u(N_{j_N(i+)}) & \text{if } u(N_i) \geq u(N_{j_N(i+)}) \\ u(N_{j_N(i-)}) & \text{otherwise} \end{cases} \mod 2.$$

Finally we denote
$$\mathfrak{J}(K) = \{j \in J(K) : K_j \text{ is not the sum of hyperbolic planes }\}.$$

1.11. Closed forms for $\theta(X(L/K))$

Throughout this section, we always assume that

(c15) $\quad\begin{cases} \dim(L^r) - \dim(K^r) \leq 2 \text{ for all } r \in \mathbb{Z}; \\ L \text{ is reducible with respect to } K \text{ completely} \end{cases}$

By Prop.1.2.7, Lemma 1.9.1 and RF I, we already have

$$\theta(X(L/K)) = \dot{F}$$

if one of the above conditions holds.

Denote $FK^{\perp}$ as the orthogonal complement of FK in FL.

First we consider one of the following cases

- either L or K is in the case of §1.4
- $s(K_i) \geq s(L_{i+1})$ for some $1 \leq i \leq s$.

Then

$$\theta(X(L/K)) \supseteq \mathfrak{u}\dot{F}^2$$

by §1.4 and Lemma 1.9.2. It is clear that one only needs to determine exactly when $\pi \in \theta(X(L/K))$. We further assume the following conditions hold

(c16) $\quad\begin{cases} mid_{K[i]}(L_K\{i-1\}) \text{ and } red_{K[i]}(L_K\{i-1\}) \text{ are even type} \\ w(L_K\{i-1\}) > s(L_K\{i-1\}) + \kappa(K[i], L_K\{i-1\}) \end{cases}$

for all $1 \leq i \leq s$. Otherwise we already have $\theta(X(L/K)) = \dot{F}$ by Lemma 1.1.4, [Xu, Prop.3] and Lemma 1.9.4.

By (c16), [Xu], Thm.1.4.1 and Thm.1.5.7, one has

(c17) $\quad \theta(SO(mid_{K[i]}(L_K\{i-1\}))) \cdot \theta(SO(red_{K[i]}(L_K\{i-1\}))) \subseteq \mathfrak{u}\dot{F}^2$

for all $1 \leq i \leq s$.

THEOREM 1.11.1. $\theta(X(L/K)) = \mathfrak{u}\dot{F}^2$ if and only if

$$\dim(L) - \dim(K) \leq 2$$

and

$$Ord(\Delta(FK^{\perp})) = 0 \quad \text{when} \quad \dim(L) - \dim(K) = 2$$

and the following conditions hold.

For any $j \in \mathfrak{J}(K)$, one denotes

$$M = red_{K[j]}(L_K\{j-1\}).$$

Let M_i be the i-th Jordan components of M.

1): If $\dim(M_1) = \dim(K_j)$, then $j_M(1) > 1$ or one of the following cases holds
   - **a):** $u(M_{j_M(2)}) - u(M) \leq 2(s(M_2) - s(M))$
   - **b):** $\dim(M_2) = 2$ and $u(M_2) = u(K[j+1]) = u(M) + 2(s(M_2) - s(M))$
   - **c):** $s(M_2) = s(K_{j+1})$ and $u(M_{j_M(2)}) - u(M) > 2(s(M_2) - s(M))$

   where $\dim(M_1) < \dim(M)$.

2): If $\dim(M_1) > \dim(K_j)$, then $j_M(1) = 1$ and

$$u(M) < u(K[j])$$

or

$$u(M) = u(K[j]) = s(K_j) \quad \text{and} \quad Ord(\Delta(FM_1)) = \infty.$$

PROOF. By (c15), (c16) and RF I, one always has
$$\theta(X(L/K)) = \theta(X(red_K(L)/K)).$$
By RF II 1), we can assume that $1 \in \mathfrak{J}(K)$. For $j = 1 \in \mathfrak{J}(K)$, one has
$$M = red_K(L).$$
If $\dim(M_1) = \dim(K_1)$, the result follows from RF II 2) and (c17) and induction.

If $\dim(M_1) > \dim(K_1)$, one can also assume that K_1 is not split by a hyperbolic plane by RF II 1). Then $j_K(1) = 1$.

Suppose $j_M(1) = 1$ and $u(M) < u(K)$. There is
$$M' = \{x \in M : Q(x) \in \mathfrak{p}\mathfrak{n}(M)\}$$
such that
$$X(M/K) = X(M'/K)$$
by RF II 3). It reduces to the above case with $j_{M'}(1) > 1$.

Suppose $j_M(1) = 1$ and
$$u(M) = u(K[j]) = s(K_j) \text{ and } Ord(\Delta(FM_1)) = \infty.$$
Then
$$\theta(X(M/K)) \subseteq \theta(X(L_K\{1\}/K[1]))\mathrm{u}\dot{F}^2.$$
by (c17). The result is obtained by induction on s.

Conversely we will show that $\pi \in \theta(X(red_K(L)/K))$ by induction on s if one of the conditions fails. By Lemma 1.1.4 and (c15), one only needs to consider that there is $j \in \mathfrak{J}(K)$ such that 1) or 2) can not be satisfied. Let j_0 be the smallest such integer. By RF I and II, one has
$$\theta(X(red_K(L)/K)) \supseteq \theta(X(red_{K[j_0]}(L_K\{j_0 - 1\})/K[j_0])) = \dot{F}.$$
The proof is complete. □

Finally we only need to deal with the case that every Jordan component of L and K is of dimension 2 and the following conditions hold

(c18) $\begin{cases} u(L_i) < s(L_i) \leq s(K_i), \ u(K_i) < s(K_i) < s(L_{i+1}) \text{ for } 1 \leq i < s \\ J(L) = \{1, \cdots, t\} \\ J(K) = \{1, \cdots, s\} \end{cases}$

by (c10). It is clear that
$$s \leq t \leq s + 1.$$

THEOREM 1.11.2. *If one of the following conditions holds*
$$\begin{cases} j_{red_{K[i]}(L\{i-1\})}(1) > 1 \\ u(L_{i+1}) < u(K_i) + 2(s(L_{i+1}) - s(K_i)) \\ u(L_{i+1}) = u(K_{i+1}) = u(K_i) + 2(s(L_{i+1}) - s(K_i)) \end{cases}$$
for $1 \leq i < t$, then
$$\theta(X(L/K))$$
$$= \theta(SO(FK^\perp)) \prod_{i=1}^{s} \theta(SO(mid_{K[i]}(L_K\{i-1\})))\theta(SO(red_{K[i]}(L_K\{i-1\}))).$$

Otherwise
$$\theta(X(L/K)) = \dot{F}.$$

PROOF. By (c15) and RF I, one has
$$X(L/K) = X(red_K(L)/K)SO(mid_K(L))$$
and $L_2 \perp \cdots \perp L_t$ splits $red_K(L)$ as the Jordan components. If $j_{red_K(L)}(1) > 1$, then
$$J(red_K(L)) = \{2, \cdots, t\}$$
by (c15), (c18) and Remark 1.7.5.

If $j_{red_K(L)}(1) = 1$ and
$$u(L_2) \leq u(K_1) + 2(s(L_2) - s(K_1)),$$
then
$$u(K_1) = u(red_K(L)).$$
We claim that
$$u(L_i) < u(K_1) + 2(s(L_i) - s(K_1))$$
for all $2 < i \leq t$.

Suppose not, there is $2 < i_0 \leq t$ such that
$$u(L_{i_0}) \geq u(K_1) + 2(s(L_{i_0}) - s(K_1)) \geq u(L_2) + 2(s(L_{i_0}) - s(L_2)).$$
A contradiction is derived by (c18) and (c10).

By the above claim, (c10) and (c18), one has
$$J(red_K(L)) = \begin{cases} J(L) & \text{if } u(L_2) < u(K_1) + 2(s(L_2) - s(K_1)) \\ J(L) \setminus \{2\} & \text{if } u(L_2) = u(K_1) + 2(s(L_2) - s(K_1)). \end{cases}$$

Therefore any Jordan decomposition
$$L_K\{1\} = M_1 \perp \cdots \perp M_{t-1}$$
satisfies
$$u(M_i) = u(L_{i+1}) \quad \text{for } 1 \leq i \leq t-1.$$
This implies that $L_K\{1\}$ and $K[2]$ satisfy the same property as that in (c18). By RF II 2), one has
$$X(red_K(L)/K) = X(L_K\{1\}/K[2])SO(red_K(L)).$$
The result follows from induction on s.

Conversely, choose the smallest integer $1 \leq i_0 \leq t-1$ such that
$$j_{red_{K[i_0]}(L\{i_0-1\})}(1) = 1$$
and one of the following holds
- $u(L_{i_0+1}) > u(K_{i_0}) + 2(s(L_{i_0+1}) - s(K_{i_0}));$
- $u(L_{i_0+1}) = u(K_{i_0}) + 2(s(L_{i_0+1}) - s(K_{i_0})) \neq u(K_{i_0+1}).$

Denote
$$N = red_{K[i_0]}(L\{i_0 - 1\}).$$
By the same argument as above inductively, there is a Jordan decomposition
$$N = N_1 \perp \cdots \perp N_{t-i_0+1}$$
such that
$$u(K_{i_0}) = u(N_1) = u(N)$$

and
$$u(N_l) = u(L_{l+i_0-1})$$
for $2 \leq l \leq t - i_0 + 1$. Therefore
$$j_N(2) > 2$$
by (c10) and
$$u(N_{j_N(2)}) - u(N) > u(N_2) - u(N_1) \geq 2(s(N_2) - s(N_1).$$
By (c15), RF I and RF II 2), one has
$$\theta(X(L/K)) \supseteq \theta(X(N/K[i_0])) = \dot{F}.$$
The proof is complete. □

CHAPTER 2

Global Theory

2.0. Introduction

In this chapter, we establish the spinor genus theory of integral representations of quadratic forms over global function fields based on the results in Chapter 1. The immediate application of this theory is to determine exactly when an integral quadratic form can be represented by an indefinite integral quadratic form with more than two variables. This chapter is organized as follows. In §2.1, we prove the realization theorem which implies that the number of spinor genera in a genus can be any power of 2. Then we study a quadratic form represented by spinor genera with codimension ≥ 2 in §2.2. In this case, either every spinor genus or exactly half of them in a genus represents the given quadratic form if this given form is represented by this genus. We also determine exactly when the exceptional case occur and which half part the spinor genus belong to. In the last section §2.3, we study codimension 0 case and prove the realization theorem which implies that every possible situation can occur. It is also given to determine when the given quadratic form is represented by the spinor genus.

Notations are standard if not explained. Let F be a global field of characteristic 2 and Ω_F be the set of all primes of F. Fix a finite non-empty set ∞ of primes of F. Let $\mathfrak{o}_F$ be the ring of all elements in F which are holomorphic outside ∞ and I_F be the idelic group of F.

Let V be a non-defective quadratic space with the associated quadratic form Q and the bilinear form $\langle\,,\,\rangle$ respectively as defined in §1.1. A quadratic lattice L in V means a finitely generated $\mathfrak{o}_F$-module in V such that FL is a non-defective quadratic space. Let

$$O(V) = \{\sigma \in GL(V) : Q(\sigma x) = Q(x) \text{ for all } x \in V\}$$

be the orthogonal group of V and $SO(V)$ the special orthogonal group which consists of all elements in $O(V)$ written as the product of the even number symmetries (see [De] and §1.1). Let $O(L)$ be the stabilizer of L by the action of $O(V)$ and $SO(L) = SO(V) \cap O(L)$.

For each $\mathfrak{p} \in \Omega$, $V_\mathfrak{p}$ (resp. $L_\mathfrak{p}$, $F_\mathfrak{p}$, etc.) denotes the local completion of V (resp. L, F, etc.). We will keep the same notations as those in the previous chapter for the local situation. For example, $\mathfrak{o}_\mathfrak{p}$ is the ring of integers in $F_\mathfrak{p}$, $\pi_\mathfrak{p}$ is the uniformizer of $F_\mathfrak{p}$ and $\mathfrak{u}_\mathfrak{p}$ is the group of units of $\mathfrak{o}_\mathfrak{p}$ for $\mathfrak{p} \in \Omega_F \setminus \infty$ and so on. We say V is indefinite if there is $\mathfrak{p} \in \infty$ such that $V_\mathfrak{p}$ is isotropic.

Let $SO_A(V)$ be the adelic group of $SO(V)$ and

$$\theta_A = \prod_{\mathfrak{p} \in \Omega} \theta_\mathfrak{p} : SO_A(V) \longrightarrow I_F/I_F^2$$

be the adelic spinor norm homomorphism. It is well-known that $SO_A(V)$ acts on $\mathfrak{o}_F$-module L of maximal rank in V. We denote $SO_A(L)$ as the stabilizer of L under such action of $SO_A(V)$.

DEFINITION 2.0.1. Two lattices of maximal rank in V are called in the same genus if they are in the same orbit of $SO_A(V)$.

Two lattices of maximal rank in V are called in the same spinor genus if they are in the same orbit of $SO(V)ker\theta_A$.

Two lattices of maximal rank in V are called in the same class if they are in the same orbit of $SO(V)$.

Set
- $gen(L) =$ the orbit of L under the action of $SO_A(V)$
- $spn(L) =$ the orbit of L under the action of $SO(V)ker\theta_A$
- $cls(L) =$ the orbit of L under the action of $SO(V)$.

Since
$$SO_A(V) \supseteq SO(V)ker\theta_A \supseteq SO(V),$$
one has
$$gen(L) \supseteq spn(L) \supseteq cls(L).$$

Let
- $h(L) =$ the number of classes in $gen(L)$
- $g(L) =$ the number of spinor genera in $gen(L)$.

The following two theorems are well-known. For completeness, we quote them here.

THEOREM 2.0.2. *For any lattice L of maximal rank, $h(L) < \infty$.*

PROOF. It follows from the reduction theory for orthogonal groups in [Ha]. A more direct and elementary proof is provided in [Co]. □

THEOREM 2.0.3. *If $\dim(V) \geq 4$ and V is indefinite, then*
$$spn(L) = cls(L)$$
for any lattice L of maximal rank in V.

PROOF. Since V is indefinite, one has that $spin(V_\infty)$ is not compact. By Strong Approximation Theorem for spin groups ([Pr]), one has
$$SO(V)ker\theta_A SO_A(L) = SO(V)SO_A(L).$$
This implies that $spn(L) = cls(L)$. □

If $\dim(V) \geq 6$, one always has that V is indefinite (see [Ri, Corollary 1.10]).

DEFINITION 2.0.4. We call
- K is represented by $gen(L)$ if there is a lattice $J \in gen(L)$ such that $K \subseteq J$.
- K is represented by $spn(L)$ if there is a lattice $J \in spn(L)$ such that $K \subseteq J$.
- K is represented by $cls(L)$ if there is a lattice $J \in cls(L)$ such that $K \subseteq J$.

For two lattices K and L in $V = FL$, we write
$$X_A(L/K) = \{\, \sigma_A \in SO_A(V) \,:\, K \subseteq \sigma_A L \,\}.$$
Then it is clear that K is represented by $gen(L)$ if and only if $X_A(L/K)$ is not empty.

For convenience, we sometimes put
$$X_\mathfrak{p}(L_\mathfrak{p}/K_\mathfrak{p}) = SO(F_\mathfrak{p} L_\mathfrak{p})$$
for $\mathfrak{p} \in \infty$, where L and K are two lattices.

We use $[L_2 : L_1]$ to denote index module ideal of two module L_2 and L_1.

2.1. Number of spinor genera in a genus

In this section we compute the number of spinor genera in a genus. By Def. 2.0.1, the number of spinor genera in $gen(L)$ is given by
$$g(L) = [SO_A(V) : SO(V)ker\theta_A SO_A(L)]$$
where $V = FL$.

If $\dim(V) \geq 4$, then
$$\theta(SO(V)) = \dot{F}$$
and one has the following isomorphism
$$\theta_A : SO_A(V)/SO(V)ker\theta_A SO_A(L) \cong I_F/\dot{F}\theta_A(SO_A(L))$$
by applying spinor norm homomorphism. Therefore
$$(c19) \qquad g(L) = [I_F : \dot{F}\theta_A(SO_A(L))]$$
(see [Co, 4.6 & 5.9]). It is clear that
$$\dot{F}\dot{F}_\infty I_F^2 \subseteq \dot{F}\theta_A(SO(L)) \subseteq I_F$$
for any $\mathfrak{o}_F$-lattice L of maximal rank in V, where
$$\dot{F}_\infty = \{(\alpha_\mathfrak{p}) \in I_F : \alpha_\mathfrak{p} = 1 \text{ for all } \mathfrak{p} \notin \infty\}.$$

LEMMA 2.1.1. *For any given $V_\mathfrak{p}$ with $\dim(V_\mathfrak{p}) \geq 4$ and any positive integer d, there exists an $\mathfrak{o}_\mathfrak{p}$-lattice L of maximal rank such that*
$$\theta_\mathfrak{p}(SO(L)) \subseteq (1+\mathfrak{p}^d)\dot{F}_\mathfrak{p}^2.$$

PROOF. It is clear that one only needs to prove the above lemma for sufficiently large d. We assume that $d > 1$.

Case I: $V_\mathfrak{p}$ *is anisotropic.*

It is clear that $\dim(V_\mathfrak{p}) = 4$ in this case and
$$V_\mathfrak{p} = (F_\mathfrak{p} e_1 + F_\mathfrak{p} f_1) \perp (F_\mathfrak{p} e_2 + F_\mathfrak{p} f_2)$$
where
$$Q(e_1) = \rho_\mathfrak{p}^{-1} Q(f_1) = \langle e_1, f_1 \rangle = 1 \text{ and } Q(e_2) = \rho_\mathfrak{p}^{-1} Q(f_2) = \langle e_2, f_2 \rangle = \pi_\mathfrak{p}$$
by [Ri, Lemma 1.7]. Then
$$V_\mathfrak{p} = W_1 \perp W_2$$
where
$$W_1 = F_\mathfrak{p} e_1 + F_\mathfrak{p}(f_1 + \pi_\mathfrak{p}^{-d-1} e_2) \text{ and } W_2 = F_\mathfrak{p} e_2 + F_\mathfrak{p}(f_2 + \pi_\mathfrak{p}^{-d} e_1)$$
and
$$Ord_\mathfrak{p}(\Delta(W_1)) = Ord_\mathfrak{p}(\Delta(W_2)) = -2d - 1.$$
Therefore
$$W_1 \cong W_2^\alpha \quad \text{for any } \alpha \notin Q(W_1).$$
By Lemma 1.1.4, one can choose
$$\alpha \in (1 + \mathfrak{p}^d).$$
Let $L_i \subset W_i$ be two lattices for $1 \leq i \leq 2$ with the norm generators
$$\pi_\mathfrak{p}^{u(L_1)} \quad \text{and} \quad \alpha\pi_\mathfrak{p}^{u(L_2)}$$

respectively and satisfying the following conditions

$$\begin{cases} u(L_1) \equiv u(L_2) \mod 2 \\ 16d < u(L_2) - u(L_1) < 2(s(L_2) - s(L_1)) \\ -8d < u(L_1) - s(L_1) < -4d - 2 \\ u(L_2) - s(L_2) < -4d - 2. \end{cases}$$

For example, one can take

$$s(L_1) = 0, \ u(L_1) = -6d, \ s(L_2) = 20d \ \text{and} \ u(L_2) = 12d.$$

Let

$$L = L_1 \perp L_2$$

as required by Thm.1.5.7 and [Xu, Prop.3].

Case II: $V_{\mathfrak{p}} = V_1 \perp \cdots \perp V_t$ where

$$V_i = F_{\mathfrak{p}} e_i + F_{\mathfrak{p}} f_i \quad \text{with} \quad \langle e_i, f_i \rangle = Q(e_i) = 1 \quad \text{and} \quad Q(f_i) = 0$$

for $1 \leq i \leq t$.

Let

$$L = L_1 \perp \cdots \perp L_t \quad \text{with} \quad L_i = \mathfrak{p}^{2di} e_i + \mathfrak{p}^{4di} f_i$$

for $1 \leq i \leq t$. Then L is as required by Thm. 1.5.7 and [Xu, Prop.3].

Case III: $V_{\mathfrak{p}}$ is in general.

By Case I and Case II, we can assume that

$$V_{\mathfrak{p}} = W \perp U$$

where W is in Case II and U is anisotropic with $2 \leq \dim(U) \leq 4$.

If $\dim(U) = 2$, we can assume that

$$d > -Ord_{\mathfrak{p}}(\Delta(U)).$$

By Lemma 1.1.4, there is $\alpha \neq 0$ such that

$$\alpha \in (1 + \mathfrak{p}^d) \cap Q(U).$$

Let $L_1 \subset U$ be an $\mathfrak{o}_{\mathfrak{p}}$-lattice such that

$$s(L_1) - u(L_1) > 2d$$

and

$$\alpha \pi_{\mathfrak{p}}^{u(L_1)} \text{ is a norm generator of } L_1.$$

Let $L_2 \subset W$ be an $\mathfrak{o}_{\mathfrak{p}}$-lattice as constructed in Case II such that

$$\theta_{\mathfrak{p}}(SO(L_2)) \subset (1 + \mathfrak{p}^d) \dot{F}_{\mathfrak{p}}^2 \quad \text{and} \quad u(L_1) \equiv u(L_2) \mod 2$$

and

$$\pi_{\mathfrak{p}}^{u(L_2)} \text{ is a norm generator of } L_2.$$

Let

$$L = L_1 \perp \mathfrak{p}^l L_2$$

for the sufficiently large integer l. Then L is as required by Thm.1.5.7 and [Xu, Prop. 3].

If $\dim(U) = 4$, there are two $\mathfrak{o}_{\mathfrak{p}}$-lattices

$$L_1 \subset U \quad \text{and} \quad L_2 \subset W$$

as constructed in Case I and Case II such that $\pi_{\mathfrak{p}}^{u(L_1)}$ and $\pi_{\mathfrak{p}}^{u(L_2)}$ are the norm generators of L_1 and L_2 respectively and
$$\theta_{\mathfrak{p}}(SO(L_1))\theta_{\mathfrak{p}}(SO(L_2)) \subseteq (1+\mathfrak{p}^d).$$
Let
$$L = L_1 \perp \mathfrak{p}^l L_2$$
for the sufficiently large l. Then L is as required by Thm.1.5.7. $\square$

By using the above lemma, we can show the following realization theorem.

THEOREM 2.1.2. *Suppose* $\dim(V) \geq 4$. *For any open subgroup* H *with*
$$\dot{F}\dot{F}_\infty I_F^2 \subseteq H \subseteq I_F,$$
there is an $\mathfrak{o}_F$-*lattice* L *of maximal rank such that*
$$\dot{F}\theta_A(SO_A(L)) = H.$$

PROOF. We prove the result in the following two steps.

Step I: *There exists an* $\mathfrak{o}_{\mathfrak{p}}$-*lattice* M *such that* $\dot{F}\theta_A(SO_A(M)) \subseteq H$.

Since H is an open subgroup in I_F, there is a finite subset S_1 of Ω_F containing ∞ such that
$$\prod_{\mathfrak{p} \in S_1} H_{\mathfrak{p}} \times \prod_{\mathfrak{p} \notin S_1} \mathfrak{u}_{\mathfrak{p}} \subseteq H$$
where $H_{\mathfrak{p}}$ is an open subgroup of $\dot{F}_{\mathfrak{p}}$ for all $\mathfrak{p} \in S_1$. Since H contains $\dot{F}\dot{F}_\infty I_F^2$, one has
$$\dot{F}(\prod_{\mathfrak{p} \in \infty} \dot{F}_{\mathfrak{p}} \times \prod_{\mathfrak{p} \in S_1 \setminus \infty} \dot{H}_{\mathfrak{p}}\dot{F}_{\mathfrak{p}}^2 \times \prod_{\mathfrak{p} \notin S_1} \mathfrak{u}_{\mathfrak{p}})I_F^2 \subseteq H.$$

For an $\mathfrak{o}_F$-lattice R of maximal rank of V, there is a finite subset S_2 of Ω_F containing ∞ such that
$$\theta_{\mathfrak{p}}(SO(R_{\mathfrak{p}})) = \mathfrak{u}_{\mathfrak{p}}\dot{F}_{\mathfrak{p}}^2 \quad \text{for any } \mathfrak{p} \notin S_2.$$
Since $H_{\mathfrak{p}}$ is open in $\dot{F}_{\mathfrak{p}}$, there exists a positive integer d such that
$$(1+\mathfrak{p}^d) \subseteq H_{\mathfrak{p}} \quad \text{for all } \mathfrak{p} \in S_1 \setminus \infty.$$
Let $S = S_1 \cup S_2$. By Lemma 2.1.1, there exists $\mathfrak{o}_{\mathfrak{p}}$-lattice $N_{\mathfrak{p}}$ such that
$$\theta_{\mathfrak{p}}(SO(N_{\mathfrak{p}})) \subseteq (1+\mathfrak{p}^d)\dot{F}_{\mathfrak{p}}^2 \quad \text{for all } \mathfrak{p} \in S \setminus \infty.$$
Then the $\mathfrak{o}_F$-lattice M satisfying
$$M_{\mathfrak{p}} = \begin{cases} N_{\mathfrak{p}} & \text{if } \mathfrak{p} \in S \setminus \infty \\ R_{\mathfrak{p}} & \text{otherwise.} \end{cases}$$
is as desired.

Step II:

Suppose
$$n = [H : \dot{F}\theta_A(SO_A(M))] > 1.$$
Then n is a positive integer by finiteness of class number of $\mathfrak{o}_F$. Let
$$T = \{\, \mathfrak{p} \in \Omega \,:\, \theta_{\mathfrak{p}}(SO(M_{\mathfrak{p}})) \neq \mathfrak{u}_{\mathfrak{p}}\dot{F}_{\mathfrak{p}}^2 \,\} \cup \infty.$$

Then T is also finite by [Xu]. By Chebotarev's Theorem (see [We, Chapter XIII, Thm.12]), there are
$$\mathfrak{q} \in \Omega \setminus T \quad \text{and} \quad i \in H \setminus \dot{F}\theta_A(SO_A(M))$$
such that
$$\mathfrak{p} - \text{component of } i = \begin{cases} \pi_{\mathfrak{p}} & \text{if } \mathfrak{p} = \mathfrak{q} \\ 1 & \text{otherwise.} \end{cases}$$
By §1.2, there exists an $\mathfrak{o}_{\mathfrak{q}}$-lattice $N_{\mathfrak{q}}$ such that
$$\theta_{\mathfrak{q}}(SO(N_{\mathfrak{q}})) = \dot{F}_{\mathfrak{q}}.$$
Let M_1 be an $\mathfrak{o}_F$-lattice such that
$$(M_1)_{\mathfrak{p}} = \begin{cases} N_{\mathfrak{q}} & \text{if } \mathfrak{p} = \mathfrak{q} \\ M_{\mathfrak{p}} & \text{otherwise.} \end{cases}$$
Then
$$i \in \dot{F}\theta_A(SO_A(M_1)) \quad \text{and} \quad [\dot{F}\theta_A(SO_A(M_1)) : \dot{F}\theta_A(SO_A(M))] = 2$$
and $\dot{F}\theta_A(SO_A(M_1)) \subseteq H$. By repeating the same argument as above to M_1, one can eventually get an $\mathfrak{o}_F$-lattice L as desired. □

COROLLARY 2.1.3. *Suppose* $\dim(V) \geq 4$.

If L is an $\mathfrak{o}_F$-lattice of maximal rank in V, then $g(L) = 2^d$ for some integer $d \geq 0$.

Conversely, for any integer $d \geq 0$, there is a $\mathfrak{o}_F$-lattice L of maximal rank such that $g(L) = 2^d$.

PROOF. It is clear that one only needs to prove the second part of Corollary 2.1.3. We will provide two proofs. The first proof is rather local and the second one is rather global.

- The first proof:
Claim: *For any given an open subgroup H with*
$$\dot{F}\dot{F}_\infty I_F^2 \subseteq H \subseteq I_F,$$
there is a proper open subgroup H' of H satisfying $\dot{F}\dot{F}_\infty I_F^2 \subseteq H'$.

Indeed, since H is open and contains $\dot{F}_\infty$, there is a finite subset $S \subset \Omega_F$ containing ∞ such that
$$I_F = \dot{F}(\prod_{\mathfrak{p} \in S} \dot{F}_{\mathfrak{p}} \times \prod_{\mathfrak{p} \notin S} \mathfrak{u}_{\mathfrak{p}})$$
and
$$\prod_{\mathfrak{p} \in \infty} \dot{F}_{\mathfrak{p}} \times \prod_{\mathfrak{p} \in S \setminus \infty} H_{\mathfrak{p}} \times \prod_{\mathfrak{p} \notin S} \mathfrak{u}_{\mathfrak{p}} \subseteq H$$
where $H_{\mathfrak{p}}$ is an open subgroup of $\mathfrak{u}_{\mathfrak{p}}$ for all $\mathfrak{p} \in S \setminus \infty$. Then for any $\mathfrak{r} \notin S$, there is $\alpha(\mathfrak{r}) \in F$ such that
$$ord_{\mathfrak{p}}(\alpha(\mathfrak{r})) = \begin{cases} 1 & \text{if } \mathfrak{p} = \mathfrak{r} \\ 0 & \text{if } \mathfrak{p} \notin S \text{ and } \mathfrak{p} \neq \mathfrak{r}. \end{cases}$$
Let
$$\mathfrak{o}_F(S) = \{x \in F : x \in \mathfrak{o}_{F_{\mathfrak{p}}} \text{ for all } \mathfrak{p} \notin S\}.$$

For any element
$$x \in \dot{F} \cap (\prod_{\mathfrak{p} \in S} \dot{F}_\mathfrak{p} \times \prod_{\mathfrak{p} \notin S} \mathfrak{u}_\mathfrak{p}) I_F^2,$$
there are integers $l_\mathfrak{r} \geq 0$ for all $\mathfrak{r} \notin S$ and almost all of them are zero such that
$$x \prod_{\mathfrak{r} \notin S} \alpha(\mathfrak{r})^{-2l_\mathfrak{r}} \in \mathfrak{o}_F(S)^\times.$$
This implies that
$$\dot{F} \cap (\prod_{\mathfrak{p} \in S} \dot{F}_\mathfrak{p} \times \prod_{\mathfrak{p} \notin S} \mathfrak{u}_\mathfrak{p}) I_F^2 = \mathfrak{o}_F(S)^\times \dot{F}^2.$$
Fix $\mathfrak{q} \in \Omega_F \setminus S$. Suppose
$$\mathfrak{u}_\mathfrak{q} = \mathfrak{o}_F(S)^\times (1 + \mathfrak{q}^n) \mathfrak{u}_\mathfrak{q}^2$$
for all positive integers n with the embedding $F \subset F_\mathfrak{q}$. Since $\mathfrak{u}_\mathfrak{q}$ and $\mathfrak{u}_\mathfrak{q}^2$ are compact, one has
$$\mathfrak{u}_\mathfrak{q} = \mathfrak{o}_F(S)^\times \mathfrak{u}_\mathfrak{q}^2.$$
The Dirichlet unit theorem ([We, Chapter IV, Thm.9]) implies that
$$\mathfrak{u}_\mathfrak{q}/\mathfrak{u}_\mathfrak{q}^2 \cong \mathfrak{o}_F(S)^\times/\mathfrak{o}_F(S)^\times \cap \mathfrak{u}_\mathfrak{q}^2$$
is finite, which is obviously not the case by [We, Chapter II, Prop.10]. There is a positive integer n_0 such that $\mathfrak{u}_\mathfrak{q}$ contains $\mathfrak{o}_F(S)^\times (1 + \mathfrak{q}^{n_0}) \mathfrak{u}_\mathfrak{q}^2$ properly. Let
$$H' = \dot{F}(\prod_{\mathfrak{p} \in \infty} \dot{F}_\mathfrak{p} \times \prod_{\mathfrak{p} \in S \setminus \infty} H_\mathfrak{p} \times (1 + \mathfrak{q}^{n_0}) \times \prod_{\mathfrak{p} \notin S \cup \{\mathfrak{q}\}} \mathfrak{u}_\mathfrak{p}) I_F^2.$$
It is clear that H' is an open subgroup of H containing $\dot{F}\dot{F}_\infty I_F^2$. Suppose
$$H' = \dot{F}(\prod_{\mathfrak{p} \in \infty} \dot{F}_\mathfrak{p} \times \prod_{\mathfrak{p} \in S \setminus \infty} H_\mathfrak{p} \times \prod_{\mathfrak{p} \notin S} \mathfrak{u}_\mathfrak{p}) I_F^2 \subseteq H.$$
One can choose an idele $i = (i_\mathfrak{p})_{\mathfrak{p} \in \Omega_F}$ as follows
$$i_\mathfrak{p} = \begin{cases} i_\mathfrak{q} \in \mathfrak{u}_\mathfrak{q} \setminus \mathfrak{o}_F(S)^\times (1 + \mathfrak{q}^{n_0}) \mathfrak{u}_\mathfrak{q}^2 & \text{if } \mathfrak{p} = \mathfrak{q} \\ 1 & \text{otherwise.} \end{cases}$$
Then $i \in H'$ and
$$i = \alpha h j^2$$
where
$$h = (h_\mathfrak{p})_{\mathfrak{p} \in \Omega_F} \in \prod_{\mathfrak{p} \in \infty} \dot{F}_\mathfrak{p} \times \prod_{\mathfrak{p} \in S \setminus \infty} H_\mathfrak{p} \times (1 + \mathfrak{q}^{n_0}) \times \prod_{\mathfrak{p} \notin S \cup \{\mathfrak{q}\}} \mathfrak{u}_\mathfrak{p}$$
and
$$\alpha \in \dot{F} \quad \text{and} \quad j = (j_\mathfrak{p})_{\mathfrak{p} \in \Omega_F} \in I_F.$$
Therefore
$$\alpha \in \dot{F} \cap (\prod_{\mathfrak{p} \in S} \dot{F}_\mathfrak{p} \times \prod_{\mathfrak{p} \notin S} \mathfrak{u}_\mathfrak{p}) I_F^2 = \mathfrak{o}_F(S)^\times \dot{F}^2.$$
Let
$$\alpha = \beta \gamma^2 \quad \text{with} \quad \beta \in \mathfrak{o}_F(S)^\times \quad \text{and} \quad \gamma \in \dot{F}.$$

Then
$$i_{\mathfrak{q}} = \beta h_{\mathfrak{q}}(\gamma j_{\mathfrak{q}})^2 \in \mathfrak{o}_F(S)^\times (1 + \mathfrak{q}^{n_0}) u_{\mathfrak{q}}^2.$$
A contradiction is derived by the choice of i. Therefore H' is contained properly in
$$\dot{F}(\prod_{\mathfrak{p} \in \infty} \dot{F}_{\mathfrak{p}} \times \prod_{\mathfrak{p} \in S \setminus \infty} H_{\mathfrak{p}} \times \prod_{\mathfrak{p} \notin S} \mathfrak{u}_{\mathfrak{p}}) I_F^2 \subseteq H$$
and the claim follows.

Let $H = I_F$. By Thm.2.1.2, there is an $\mathfrak{o}_F$-lattice L such that
$$\dot{F}\theta_A(SO_A(L_0)) = I_F.$$
Therefore $g(L) = 1$ by (c19). By the claim, there is an open subgroup H' such that H/H' is a finite elementary abelian 2-group. Let H_1 be a subgroup H such that
$$\dot{F}\dot{F}_\infty I_F^2 \subseteq H' \subseteq H_1 \subseteq H \quad \text{and} \quad [H : H_1] = 2.$$
Then H_1 is open as well. By Thm. 2.1.2, there is an $\mathfrak{o}_F$-lattice such that
$$\dot{F}\theta_A(SO_A(L_1)) = H_1.$$
Therefore $g(L_1) = 2$ by (c19). By applying the claim to H_1, there is an open subgroup H_2 such that
$$\dot{F}\dot{F}_\infty I_F^2 \subseteq H_2 \subseteq H_1 \quad \text{and} \quad [H_1 : H_2] = 2$$
and there is an $\mathfrak{o}_F$-lattice L_2 such that
$$\dot{F}\theta_A(SO_A(L_2)) = H_2.$$
Therefore $g(L_2) = 2^2$ by (c19). The result follows from repeating the same argument.

- The second proof:

Let $\mathfrak{q}_1, \cdots, \mathfrak{q}_d \in \Omega_F \setminus \infty$. By Chinese Remainder Theorem, there are $\alpha_1, \cdots, \alpha_d \in F$ such that
$$\alpha_i - \pi_{\mathfrak{q}_i}^{-1} \in \mathfrak{q}_i \quad \text{and} \quad \alpha_i \in \mathfrak{p} \cap \mathfrak{q}_j \quad \text{for all } \mathfrak{p} \in \infty \text{ and } j \neq i$$
for $1 \leq i \leq d$. Then the polynomials
$$x^2 + x + \alpha_i \quad \text{for } 1 \leq i \leq d$$
are split in $F_{\mathfrak{p}}$ and $F_{\mathfrak{q}_j}$ for all $\mathfrak{p} \in \infty$ and $j \neq i$ by Hensel's Lemma and irreducible over $F_{\mathfrak{q}_i}$. Let E be the splitting field of all these polynomials over F. Then E/F is a Galois extension with
$$[E : F] = 2^d \quad \text{and} \quad Gal(E/F) \cong \mathbb{Z}/2 \times \cdots \times \mathbb{Z}/2.$$
By classfield theory (see [We1]), one has $\dot{F}N_{E/F}(I_E)$ is an open subgroup of I_F and
$$I_F/\dot{F}N_{E/F}(I_E) \cong Gal(E/F) \cong \mathbb{Z}/2 \times \cdots \times \mathbb{Z}/2.$$
Therefore
$$I_F^2 \subseteq \dot{F}N_{E/F}(I_E).$$
Since ∞ splits in E completely, one has
$$\dot{F}_\infty \subseteq \dot{F}N_{E/F}(I_E).$$

Therefore $\dot{F}N_{E/F}(I_E)$ is an open subgroup of I_F such that
$$\dot{F}\dot{F}_\infty I_F^2 \subseteq \dot{F}N_{E/F}(I_E) \subseteq I_F.$$
By Thm.2.1.2, there is an $\mathfrak{o}_F$-lattice L such that
$$\dot{F}\theta_A(SO_A(L)) = \dot{F}N_{E/F}(I_E).$$
Therefore $g(L) = 2^d$ by (c19). $\square$

EXAMPLE 2.1.4. Suppose
$$F = \mathbb{F}_2(t), \quad \infty = (\frac{1}{t}) \quad \text{and} \quad \mathfrak{o}_F = \mathbb{F}_2[t].$$
Let L be an $\mathbb{F}_2[t]$-lattice for any given integer $d \geq 0$ such that the corresponding quadratic form is
$$f(x,y,z,w) = x^2 + t^m xy + t^{2m} y^2 + t^m(z^2 + t^m zw + t^{2m} w^2),$$
where $m = 2(2d+1)$. Then $h(L) = 2^d$.

PROOF. By [Xu, Prop.1 and Prop.3] and Thm.1.5.7, we have
$$\mathfrak{p} - \text{component of } \theta_A(SO_A(L)) = \begin{cases} (1+\mathfrak{p}^{2d+1})\dot{F}_\mathfrak{p}^2 & \text{if } \mathfrak{p} = (t) \\ \dot{F}_\mathfrak{p} & \text{if } \mathfrak{p} = \infty \\ \mathfrak{u}_\mathfrak{p} \dot{F}_\mathfrak{p}^2 & \text{otherwise.} \end{cases}$$

Since
$$\mathfrak{u}_\mathfrak{p} \dot{F}_\mathfrak{p}^2 / (1+\mathfrak{p}^{2d+1})\dot{F}_\mathfrak{p}^2 \cong (1+\mathfrak{p})\dot{F}_\mathfrak{p}^2 / (1+\mathfrak{p}^{2d+1})\dot{F}_\mathfrak{p}^2$$
is a $\mathbb{Z}/2\mathbb{Z}$-vector space with the basis
$$\{1 + t^i : i = 1, 3, \cdots, 2d-1\}$$
for $\mathfrak{p} = (t)$, one has
$$[\mathfrak{u}_\mathfrak{p} \dot{F}_\mathfrak{p}^2 : (1+\mathfrak{p}^{2d+1})\dot{F}_\mathfrak{p}^2] = 2^d.$$
Since the class number of F is one, we have
$$I_F = \dot{F}(\dot{F}_\infty \times \prod_{\mathfrak{p}<\infty} \mathfrak{u}_\mathfrak{p}) = \dot{F}(\dot{F}_\infty \times \prod_{\mathfrak{p}<\infty} \mathfrak{u}_\mathfrak{p})I_F^2$$
and
$$\dot{F} \cap (\dot{F}_\infty \times \prod_{\mathfrak{p}<\infty} \mathfrak{u}_\mathfrak{p})I_F^2 = \dot{F} \cap (\dot{F}_\infty \times (1+(t)^{2d+1}) \times \prod_{\mathfrak{p}<\infty, \mathfrak{p}\neq(t)} \mathfrak{u}_\mathfrak{p})I_F^2 = \dot{F}^2.$$
Since
$$\frac{\dot{F}(\dot{F}_\infty \times \prod_{\mathfrak{p}<\infty}\mathfrak{u}_\mathfrak{p})I_F^2}{\dot{F}(\dot{F}_\infty \times (1+(t)^{2d+1}) \times \prod_{\mathfrak{p}<\infty,\mathfrak{p}\neq(t)} \mathfrak{u}_\mathfrak{p})I_F^2}$$
$$\cong \left(\frac{(\dot{F}_\infty \times \prod_{\mathfrak{p}<\infty}\mathfrak{u}_\mathfrak{p})I_F^2}{\dot{F}\cap(\dot{F}_\infty \times \prod_{\mathfrak{p}<\infty}\mathfrak{u}_\mathfrak{p})I_F^2}\right) / \left(\frac{(\dot{F}_\infty \times (1+(t)^{2d+1}) \times \prod_{\mathfrak{p}<\infty,\mathfrak{p}\neq(t)}\mathfrak{u}_\mathfrak{p})I_F^2}{\dot{F}\cap(\dot{F}_\infty \times (1+(t)^{2d+1}) \times \prod_{\mathfrak{p}<\infty,\mathfrak{p}\neq(t)}\mathfrak{u}_\mathfrak{p})I_F^2}\right)$$
$$\cong \frac{(\dot{F}_\infty \times \prod_{\mathfrak{p}<\infty}\mathfrak{u}_\mathfrak{p})I_F^2}{(\dot{F}_\infty \times (1+(t)^{2d+1}) \times \prod_{\mathfrak{p}<\infty}\mathfrak{u}_\mathfrak{p})I_F^2} \cong \frac{\mathfrak{u}_{(t)}\dot{F}_{(t)}^2}{(1+(t)^{2d+1})\dot{F}_{(t)}^2},$$

we have $g(L) = 2^d$ by (c19). Since
$$f(t^{\frac{m}{2}}, 0, 1, 0) = 0,$$
one has $h(L) = g(L)$ by Thm.2.0.3. □

2.2. Representations of spinor genera, codimension ≥ 2

Let L be an $\mathfrak{o}_F$-lattice in $V = FL$ and $\dim(L) = m \geq 4$. Suppose K is represented by $gen(L)$. Then there exists

$$\sigma \in SO_A(V) \text{ such that } K \subseteq \sigma L.$$

By Thm.1.6.1, we have

$$\theta_A(X_A(\sigma L/K)) \text{ is a group.}$$

For any $\tau \in SO_A(V)$ with $K \subseteq \tau L$, one has

$$X_A(\tau L/K)\tau = X_A(\sigma L/K)\sigma.$$

This implies that this group is independent of the choice of σ.

DEFINITION 2.2.1. We denote

$$\theta_A(gen(L) : K) = \theta_A(X(\sigma L/K)).$$

It is clear that

(c20) $$\theta_A(SO_A(FK^\perp)) \subseteq \theta_A(gen(L) : K).$$

A natural question is to determine how many spinor genera in $gen(L)$ which represent K. Let $\lambda(gen(L), K)$ denote such number.

There is a natural one to one correspondence as follows

$$\{ \text{ spinor genera in } gen(L) \text{ representing } K \}$$
$$\longleftrightarrow X_A(\sigma L/K)SO(V)ker\theta_A/SO_A(\sigma L)SO(V)ker\theta_A.$$

Since there is an isomorphism induced by θ_A

$$X_A(\sigma L/K)SO(V)ker\theta_A/SO_A(\sigma L)SO(V)ker\theta_A$$
$$\cong \dot{F}\theta_A(gen(L) : K)/\dot{F}\theta_A(SO_A(L))$$

by Thm.1.6.1, Def.2.2.1 and [Co, 3.6], one has

(c21) $$\lambda(gen(L), K) = [\dot{F}\theta_A(gen(L) : K) : \dot{F}\theta_A(SO_A(L))].$$

Let

$$\delta = \dim(L) - \dim(K).$$

PROPOSITION 2.2.2. *If $\delta \geq 4$ or $\delta = 2$ and $FK^\perp$ is isotropic, then every spinor genus in $gen(L)$ represents K.*

PROOF. Since

$$\theta_A(SO_A(FK^\perp)) = I_F$$

if $\delta \geq 4$ or $\delta = 2$ and $FK^\perp$ is isotropic, one has

$$\theta_A(gen(L) : K) = I_F$$

by (c20). Therefore

$$\lambda(gen(L) : K) = g(L)$$

by (c19) and (c21). □

2.2. REPRESENTATIONS OF SPINOR GENERA, CODIMENSION ≥ 2

For the rest of this section, we will consider
$$\delta = 2 \quad \text{and} \quad FK^\perp \text{ is anisotropic.}$$
Let ζ be in an algebraic closure of F such that
$$\zeta^2 + \zeta \in \Delta(FK^\perp) \quad \text{and} \quad E = F(\zeta).$$
Then
$$[E:F] = 2 \quad \text{and} \quad \theta_A(SO_A(FK^\perp)) = N_{E/F}(I_E).$$
By the class field theory, one has
$$[I_F : \dot{F} N_{E/F}(I_E)] = [I_F : \dot{F}\theta_A(SO_A(FK^\perp))] = 2.$$
Therefore
$$\lambda(gen(L):K) = \frac{1}{2}g(L) \quad \text{or} \quad \lambda(gen(L):K) = g(L)$$
by (c20) and (c21).

DEFINITION 2.2.3. K is called a spinor exception for $gen(L)$ if K is codimension 2 in L and
$$\lambda(gen(L):K) = \frac{1}{2}g(L).$$

By the following theorem, we will find that it is a purely local question to determine if K is a spinor exception for $gen(L)$.

THEOREM 2.2.4. *Suppose K is the codimension 2 in L. K is a spinor exception for $gen(L)$ if and only if*
$$\theta_A(gen(L):K) = N_{E/F}(I_E).$$

PROOF. By (c20), one only needs to prove that
$$\theta_A(gen(L):K) \subseteq N_{E/F}(I_E).$$
Suppose
$$\theta_A(gen(L):K) \not\subseteq N_{E/F}(I_E).$$
Then there are
$$(i_\mathfrak{p})_{\mathfrak{p} \in \Omega_F} \in \theta_A(gen(L):K) \setminus N_{E/F}(I_E)$$
and $\mathfrak{q} \in \Omega_F$ such that
$$i_\mathfrak{p} = 1 \quad \text{for all } \mathfrak{p} \in \Omega_F \text{ except } \mathfrak{q}$$
and
$$i_\mathfrak{q} \notin G((F\mathfrak{q}K)^\perp).$$
Therefore $\mathfrak{q}$ is ramified or inert in E/F. Let $\mathfrak{Q}$ be the prime in E above $\mathfrak{q}$. Then
$$Gal(E/F) = Gal(E_\mathfrak{Q}/F_\mathfrak{q}).$$
Let
$$\chi \text{ be the non-trivial character of } Gal(E/F)$$
and
$$\varrho = \prod_{\mathfrak{p} \in \Omega_F} \varrho_\mathfrak{p} \text{ be the Artin map of } E/F$$
where $\varrho_\mathfrak{p}$ is the local Artin map of $F_\mathfrak{p}$ for all $\mathfrak{p} \in \Omega_F$. Then
$$\chi(\varrho_\mathfrak{q}(i_\mathfrak{q})) = -1$$
by the local classfield theory [We, Chapter XII, Thm.4].

By Def. 2.2.3 and (c19) and (c21), one has
$$\dot{F}\theta_A(gen(L):K) = \dot{F}N_{E/F}(I_E).$$
There is $\alpha \in \dot{F}$ such that
$$\alpha(i_\mathfrak{p})_{\mathfrak{p}\in\Omega_F} \in N_{E/F}(I_E).$$
Then
$$\prod_{\mathfrak{p}\in\Omega_F} \chi(\varrho_\mathfrak{p}(\alpha i_\mathfrak{p})) = 1$$
by [We, Chapter XII, Thm.4].

By Artin Reciprocity Law [We, Chapter XIII, Thm. 2, Corollary], one has
$$\prod_{\mathfrak{p}\in\Omega_F} \chi(\varrho_\mathfrak{p}(\alpha)) = 1.$$
Therefore
$$\prod_{\mathfrak{p}\in\Omega_F} \chi(\varrho_\mathfrak{p}(i_\mathfrak{p})) = \chi(\varrho_\mathfrak{q}(i_\mathfrak{q})) = 1.$$
A contradiction is derived. $\square$

Suppose K is a spinor exception for $gen(L)$. Then there are exactly half of spinor genera in $gen(L)$ representing K. For a lattice in $gen(L)$, one needs to determine which half of $gen(L)$ this lattice belongs to.

The following proposition gives the relation between the index module ideals and the spinor norms.

PROPOSITION 2.2.5. *Suppose $L_\mathfrak{p}$ is unimodular and $n(L_\mathfrak{p}) = n(K_\mathfrak{p})$ for some $\mathfrak{p} \in \Omega_F \setminus \infty$.*
If $[K_\mathfrak{p} : L_\mathfrak{p} \cap K_\mathfrak{p}] = \mathfrak{p}^r$, then
$$\theta_\mathfrak{p}(X(L_\mathfrak{p}/K_\mathfrak{p})) = \pi_\mathfrak{p}^r \mathfrak{u}_\mathfrak{p} \dot{F}_\mathfrak{p}^2 \quad or \quad \dot{F}_\mathfrak{p}.$$

PROOF. By [Xu, Prop.1 and 2] and Thm. 1.6.1, the proposition is true for $r = 0$. Assume that $r \geq 1$. By [Xu, Prop.2], one can further assume that
$$L_\mathfrak{p} = M \perp (\mathfrak{o}_\mathfrak{p} y + \mathfrak{o}_\mathfrak{p} z)$$
where M is the orthogonal sum of hyperbolic planes and
$$Q(y)\mathfrak{o}_\mathfrak{p} = \mathfrak{n}(L_\mathfrak{p}), \ \ ord_\mathfrak{p}(Q(y)) \equiv 0 \mod 2, \ \ ord_\mathfrak{p}(Q(z)) \geq 0 \ \text{ and } \ \langle y, z \rangle = 1.$$
Let
$$\langle L_\mathfrak{p}, K_\mathfrak{p} \rangle = \mathfrak{p}^{-d}.$$
Then d is a positive integer. There is a vector $e \in K_\mathfrak{p}$ such that
$$\langle e, L_\mathfrak{p} \rangle = \mathfrak{p}^{-d}.$$
One can assume that
$$Q(e)\mathfrak{o}_\mathfrak{p} = \mathfrak{n}(K_\mathfrak{p}).$$
Otherwise there is $e' \in K$ satisfying $Q(e')\mathfrak{o}_\mathfrak{p} = \mathfrak{n}(K_\mathfrak{p})$ and $\langle e', L \rangle$ is a proper subset of $\mathfrak{p}^{-d}$. Then we can replace e by $e + e'$ as required.

Claim: *There is $x \in L_\mathfrak{p}$ such that*
$$Q(x) = Q(e) \quad and \quad \langle x, e \rangle \mathfrak{o}_\mathfrak{p} = \mathfrak{p}^{-d}.$$

Indeed, there are $\alpha \in \mathfrak{u}_\mathfrak{p}$, $\beta \in \mathfrak{o}_\mathfrak{p}$ and $w \in M$ such that
$$Q(e) = Q(\alpha y + \beta z + w)$$
since $K_\mathfrak{p}$ is represented by $L_\mathfrak{p}$ and $\mathfrak{n}(K_\mathfrak{p}) = \mathfrak{n}(L_\mathfrak{p})$. Then
$$Q(e) - \alpha^2 Q(y) \in \mathfrak{o}_\mathfrak{p}.$$
By Def.1.1.1, there is $\alpha_1 \in \mathfrak{u}_\mathfrak{p}$ such that
$$Q(e) - \alpha_1^2 Q(y) \in \mathfrak{p}.$$
By Hensel's lemma, there is $\beta_1 \in \mathfrak{p}$ such that
$$Q(e) = \alpha_1^2 Q(y) + \alpha_1 \beta_1 + \beta_1^2 Q(z) = Q(\alpha_1 y + \beta_1 z).$$
If $\langle e, y \rangle \mathfrak{o}_\mathfrak{p} = \mathfrak{p}^{-d}$, then one can choose
$$x = \alpha_1 y + \beta_1 z$$
as required.

If $\langle e, y \rangle \mathfrak{o}_\mathfrak{p}$ is contained in $\mathfrak{p}^{-d}$ properly and $\langle e, z \rangle \mathfrak{o}_\mathfrak{p} = \mathfrak{p}^{-d}$, then one can replace y by $y + z$ when
$$ord_\mathfrak{p}(Q(y)) < ord_\mathfrak{p}(Q(z))$$
or interchange y and z when
$$ord_\mathfrak{p}(Q(y)) = ord_\mathfrak{p}(Q(z))$$
and repeat the above argument.

Otherwise there is a vector $v \in M$ such that
$$Q(v) = 0 \quad \text{and} \quad \langle v, e \rangle \mathfrak{o}_\mathfrak{p} = \mathfrak{p}^{-d}$$
since
$$\langle e, L_\mathfrak{p} \rangle = \langle e, y \rangle \mathfrak{o}_\mathfrak{p} + \langle e, z \rangle \mathfrak{o}_\mathfrak{p} + \langle e, M \rangle = \langle e, M \rangle = \mathfrak{p}^{-d}$$
and M is the orthogonal sum of hyperbolic planes. Let
$$x = \alpha_1 y + \beta_1 z + v$$
as required and the claim follows.

By the above claim, there is a sublattice L' of $L_\mathfrak{p}$ such that
$$L_\mathfrak{p} = L' \perp (\mathfrak{p}^d e + \mathfrak{o}_\mathfrak{p} x).$$
Let
$$\tau = \tau_{e+x}.$$
Then
$$\tau(L_\mathfrak{p}) = L' \perp (\mathfrak{o}_\mathfrak{p} e + \mathfrak{p}^d x).$$
For any
$$a\pi_\mathfrak{p}^d e + bx + l \in L_\mathfrak{p} \cap K_\mathfrak{p}$$
with $a, b \in \mathfrak{o}_\mathfrak{p}$ and $l \in L'$, we have
$$b\langle x, e \rangle = \langle p^d e + bx + c, e \rangle \in \mathfrak{s}(K_\mathfrak{p}) \subseteq \mathfrak{o}_\mathfrak{p}.$$
Then
$$b \in \mathfrak{p}^d$$
by the above claim. Therefore
$$K_\mathfrak{p} \cap L_\mathfrak{p} = K_\mathfrak{p} \cap \tau(L_\mathfrak{p}) \cap L_\mathfrak{p}$$

and
$$[K_\mathfrak{p} : K_\mathfrak{p} \cap L_\mathfrak{p}] = [K_\mathfrak{p} : K_\mathfrak{p} \cap \tau(L_\mathfrak{p})][K_\mathfrak{p} \cap \tau(L_\mathfrak{p}) : K_\mathfrak{p} \cap \tau(L_\mathfrak{p}) \cap L_\mathfrak{p}].$$

There is an $\mathfrak{o}_\mathfrak{p}$-module isomorphism
$$\frac{K_\mathfrak{p} \cap \tau(L_\mathfrak{p})}{K_\mathfrak{p} \cap \tau(L_\mathfrak{p}) \cap L_\mathfrak{p}} \cong \frac{L_\mathfrak{p} + (K_\mathfrak{p} \cap \tau(L_\mathfrak{p}))}{L_\mathfrak{p}}.$$

Since
$$L_\mathfrak{p} + (K_\mathfrak{p} \cap \tau(L_\mathfrak{p})) = L' \perp (\mathfrak{o}_\mathfrak{p} e + \mathfrak{o}_\mathfrak{p} x),$$
one has
$$[L_\mathfrak{p} + (K_\mathfrak{p} \cap \tau(L_\mathfrak{p})) : L_\mathfrak{p}] = \mathfrak{p}^d.$$

Therefore
$$[K_\mathfrak{p} : K_\mathfrak{p} \cap \tau(L_\mathfrak{p})] = \mathfrak{p}^{r-d}.$$

It is clear that
$$\theta_\mathfrak{p}(\tau) \in \pi_\mathfrak{p}^d \mathfrak{u}_\mathfrak{p} \dot{F}_\mathfrak{p}^2.$$

The result follows from replacing $L_\mathfrak{p}$ by $\tau(L_\mathfrak{p})$ and induction on r. □

EXAMPLE 2.2.6. The condition of $n(L_\mathfrak{p}) = n(K_\mathfrak{p})$ in Prop. 2.2.5 can not be removed.

Suppose d and s are positive integers and $d > s$. Let
$$L_\mathfrak{p} = (\mathfrak{o}_\mathfrak{p} e + \mathfrak{o}_\mathfrak{p} f) \perp (\mathfrak{o}_\mathfrak{p} x + \mathfrak{o}_\mathfrak{p} y)$$
where
$$Q(e) = Q(f) = 0 \quad \text{and} \quad \langle e, f \rangle = 1$$
and
$$Q(x) = \pi_\mathfrak{p}^{-2d}, \quad Q(y) = \rho_\mathfrak{p} \pi_\mathfrak{p}^{2d} \quad \text{and} \quad \langle x, y \rangle = 1.$$
Let
$$w = e + \pi_\mathfrak{p}^d x \quad \text{and} \quad v = \frac{1}{1 + \rho_\mathfrak{p} \pi_\mathfrak{p}^s}(\pi_\mathfrak{p}^{-s} e + \rho_\mathfrak{p} \pi_\mathfrak{p}^s f + \pi_\mathfrak{p}^{-d} y)$$
and
$$K_\mathfrak{p} = \mathfrak{o}_\mathfrak{p} w + \mathfrak{o}_\mathfrak{p} v.$$

Since
$$\tau_{v+f}(w) = w \in L_\mathfrak{p} \quad \text{and} \quad \tau_{v+f}(v) = f \in L_\mathfrak{p},$$
one has
$$\theta_\mathfrak{p}(X(L_\mathfrak{p}/K_\mathfrak{p})) = \theta_\mathfrak{p}(\tau_{v+f}) \mathfrak{u}_\mathfrak{p} \dot{F}_\mathfrak{p}^2 = \pi_\mathfrak{p}^s \mathfrak{u}_\mathfrak{p} \dot{F}_\mathfrak{p}^2$$
by RF II 1) and [Xu, Prop.2].

Since
$$K_\mathfrak{p} \cap L_\mathfrak{p} = \mathfrak{o}_\mathfrak{p} w + \mathfrak{p}^d v,$$
one has
$$[K_\mathfrak{p} : L_\mathfrak{p} \cap K_\mathfrak{p}] = \mathfrak{p}^d.$$
□

Let
$$S = \{\mathfrak{p} \in \Omega_F \setminus \infty : \mathfrak{p} | \mathfrak{v}(L) \text{ or } n(L_\mathfrak{p}) \neq n(K_\mathfrak{p})\} \cup \{\infty\}.$$
Then S is finite. By the Weak Approximation Theorem, there is $M \in cls(L)$ such that
(c22) $$K_\mathfrak{p} \subseteq M_\mathfrak{p} \quad \text{for all} \quad \mathfrak{p} \in S.$$

Let
$$\Im(L,K) = \{[K : M \cap K] \;:\; M \in cls(L) \text{ satisfying } (c22)\,\}.$$

THEOREM 2.2.7. *Suppose K is a spinor exception for $gen(L)$. Then K is represented by $spn(L)$ if and only if there is*
$$\mathfrak{a} \in \Im(L,K)$$
such that $\mathfrak{a}$ is trivial in $Gal(E/F)$ under the Artin map.

PROOF. Let $M \in cls(L)$ satisfying (c22) such that
$$\mathfrak{a} = [K : K \cap M] = \prod_{\mathfrak{p} \notin S} \mathfrak{p}^{d_\mathfrak{p}}.$$

Then $d_\mathfrak{p} = 0$ for almost all $\mathfrak{p} \notin S$ and every $\mathfrak{p} \notin S$ is unramified in E/F by Thm. 2.2.4, [Xu, Prop.1 and 2] and [We, Chapter XIII, Prop.14, Corollary 3].

Since K is represented by $gen(L)$, there is
$$\tau = (\tau_\mathfrak{p})_{\mathfrak{p} \in \Omega_F} \in SO_A(FL) \quad \text{such that} \quad K \subseteq \tau(M).$$
By (c22), one can assume that
$$\tau_\mathfrak{p} = 1 \quad \text{for } \mathfrak{p} \in S.$$
If
$$\theta_\mathfrak{p}(X(L_\mathfrak{p}/K_\mathfrak{p})) = \dot{F}_\mathfrak{p}$$
for $\mathfrak{p} \notin S$, then $\mathfrak{p}$ splits completely in E/F by [We, Chapter XIII, Prop.14 Corollary 2] and the $Frob(\mathfrak{p})$ is trivial in $Gal(E/F)$.

Otherwise one has
$$\theta_\mathfrak{p}(\tau_\mathfrak{p}) = \epsilon \pi_\mathfrak{p}^{d_\mathfrak{p}} \quad \text{with} \quad \epsilon \in \mathfrak{u}_\mathfrak{p}$$
for $\mathfrak{p} \notin S$ in this case by Prop.2.2.5.

Therefore
$$\varrho(\theta_A(\tau)) = \prod_{\mathfrak{p} \notin S}(Frob(\mathfrak{p}))^{d_\mathfrak{p}}$$
where ϱ is the Artin map of E/F. One concludes that

$\mathfrak{a}$ is trivial in $Gal(E/F)$ by Artin map
$$\Leftrightarrow \theta_A(\tau) \in \dot{F} N_{E/F}(I_E) = \dot{F}\theta_A(gen(L) : K)$$
$$\Leftrightarrow K \text{ is represented by } spn(M) = spn(L)$$

by Thm.2.2.4. $\square$

Write
$$FL = FK \perp W$$
and
$$\Upsilon(L,K) = \{\,\mathfrak{p} \in \Omega_F \setminus \infty \,:\, \theta_\mathfrak{p}(X(L_\mathfrak{p}/K_\mathfrak{p})) \neq \theta_\mathfrak{p}(SO(W_\mathfrak{p}))\}.$$
It is clear that $\Upsilon(L,K)$ is finite. We use $|\Upsilon(L,K)|$ to denote the number of elements in $\Upsilon(L,K)$.

The following result is a variant version of Thm.2.2.7.

THEOREM 2.2.8. *Suppose K is a spinor exception for $gen(L)$. Then K is represented by $spn(L)$ if and only if $|\Upsilon(L,K)|$ is even.*

PROOF. Since K is represented by $gen(L)$, there is
$$\sigma_A = (\sigma_{\mathfrak{p}})_{\mathfrak{p}} \in \Omega_F \in SO_A(FL) \quad \text{such that} \quad K \subseteq \sigma_A(L).$$
Then K is represented by $spn(L)$ if and only if
$$\theta_A(\sigma_A) \in \dot{F}\theta_A(gen(L):K) = \dot{F}N_{E/F}(I_E)$$
by Thm.2.2.4. Let χ be the non-trivial character of $Gal(E/F)$ and ϱ be the Artin map of E/F. Then
$$\chi(\varrho(\theta_{\mathfrak{p}}(\sigma_{\mathfrak{p}}))) = -1 \Leftrightarrow \mathfrak{p} \in \Upsilon(L,K).$$
Since
$$\theta_A(\sigma_A) \in \dot{F}N_{E/F}(I_E) \Leftrightarrow \chi(\varrho(\theta_A(\sigma_A))) = 1 = \prod_{\mathfrak{p} \in \Omega_F} \chi(\varrho(\theta_{\mathfrak{p}}(\sigma_{\mathfrak{p}}))),$$
the result follows. $\square$

Now we give infinitely many examples so that the local-global principle does not hold for integral representations of quadratic forms in characteristic 2.

EXAMPLE 2.2.9. Suppose
$$F = \mathbb{F}_2(t), \quad \infty = (\frac{1}{t}) \quad \text{and} \quad \mathfrak{o}_F = \mathbb{F}_2[t].$$
For any positive integer d, let $m = 2(2d+1)$ and
$$\varphi(x,y,z,w) = x^2 + t^m xy + t^{2m} y^2 + t^m z^2 + t^{2m} zw + t^{3m} w^2$$
and
$$\psi(x,y) = t^{2m-2} x^2 + t^{6d+2} xy + t^{m-2} y^2.$$
Then ψ is represented by $gen(\varphi)$ but not φ itself.

PROOF. Let
$$L = \mathfrak{o}_F e_1 + \mathfrak{o}_F f_1 + \mathfrak{o}_F e_2 + \mathfrak{o}_F f_2$$
be an $\mathfrak{o}_F$-lattice such that the associated quadratic form
$$Q(xe_1 + yf_1 + ze_2 + wf_2) = \varphi(x,y,z,w).$$
Let
$$K = \mathfrak{o}_F e + \mathfrak{o}_F f$$
where
$$e = (1+t)^{-1}(t^{m-2} e_1 + t^{2d-1} e_2 + f_1) \quad \text{and} \quad f = (1+t)t^{2d} e_1 + e_2.$$
Then the associated quadratic form of K is
$$Q(xe + yf) = \psi(x,y).$$
It is clear that
$$K_{\mathfrak{p}} \subset L_{\mathfrak{p}} \quad \text{for} \quad \mathfrak{p} \in \Omega_F \setminus \{1+t, \infty\}.$$
For $\mathfrak{p} = 1+t$, one has
$$\tau_{e+t^{m-1}e_1} e = t^{m-1} e_1 \in L_{\mathfrak{p}} \quad \text{and} \quad \tau_{e+t^{m-1}e_1} f \in L_{\mathfrak{p}}.$$
Therefore K is represented by $gen(L)$.
Write
$$FL = FK \perp W.$$
Then
$$\Delta(W) = \Delta(FL) + \Delta(FK) = \Delta(FK) = t^{-2}.$$

2.2. REPRESENTATIONS OF SPINOR GENERA, CODIMENSION ≥ 2

Since
$$t^{-2} \in \wp(F_\infty) \quad \text{and} \quad t^{-2} \notin \wp(F_{(t)}),$$
one has
$$\Delta(W) \in \wp(F_\infty) \quad \text{and} \quad \Delta(W) \notin \wp(F_{(t)}).$$
Therefore, one has
$$\theta_\infty(SO(W_\infty)) = \dot{F}_\infty = \theta_\infty(X(L_\infty/K_\infty))$$
and W is anisotropic.

If
$$\mathfrak{p} \notin \{\infty, 1+t, t\},$$
then
$$\mathfrak{s}(L_\mathfrak{p}) = \mathfrak{s}(K_\mathfrak{p}).$$
Since
$$Ord_\mathfrak{p}(\Delta(W)) \geq 0$$
in this case, one has
$$\theta_\mathfrak{p}(SO(L_\mathfrak{p})) = \mathfrak{u}_\mathfrak{p} \dot{F}_\mathfrak{p}^2 \subseteq \theta_\mathfrak{p}(SO(W_\mathfrak{p}))$$
by [Xu, Prop. 1] and Lemma 1.1.4. Since $K_\mathfrak{p}$ and $L_\mathfrak{p}$ are not of W-type, one has
$$\theta_\mathfrak{p}(X(L_\mathfrak{p}/K_\mathfrak{p})) = \theta_\mathfrak{p}(SO(W_\mathfrak{p}))$$
for $(\mathfrak{p}, (1+t)t) = 1$ by RF II.

If $\mathfrak{p} = t$, then
$$Ord_\mathfrak{p}(\Delta(W)) = -1.$$
By Thm.1.5.7 and Lemma 1.1.4, one has
$$\theta_\mathfrak{p}(SO(L_\mathfrak{p})) = (1 + \mathfrak{p}^{2d+1}) \dot{F}_\mathfrak{p}^2 \subseteq \theta_\mathfrak{p}(SO(W_\mathfrak{p})).$$
By RF I and RF II 2) i), one has
$$\theta_\mathfrak{p}(X(L_\mathfrak{p}/K_\mathfrak{p})) = \theta_\mathfrak{p}(SO(W_\mathfrak{p})).$$

If $\mathfrak{p} = 1 + t$, then
$$Ord_\mathfrak{p}(\Delta(W)) = 0$$
and
$$\theta_\mathfrak{p}(X(\tau_{e+t^{m-1}e_1} L_\mathfrak{p}/K_\mathfrak{p})) = \theta_\mathfrak{p}(SO(W_\mathfrak{p})) = \mathfrak{u}_\mathfrak{p} \dot{F}_\mathfrak{p}^2$$
by [Xu, Prop.1] and RF II 3) ii).

Therefore K is a spinor exception for $gen(L)$ by Thm.2.2.4.
Since
$$\theta(\tau_{e+t^{m-1}e_1}) = t^{2m-1}(1+t)^{-1}\dot{F}^2,$$
one has
$$\Upsilon(L, K) = \{1+t\}$$
and the result follows from Thm.2.2.8. $\square$

2.3. Representations of spinor genera, codimension 0

In this section, we study the remain case that $\dim(K) = \dim(L)$. First one has the following realization theorem.

THEOREM 2.3.1. *Suppose L is an $\mathfrak{o}_F$-lattice with the maximal rank ≥ 4. For any subgroup H with*

$$SO_A(V) \supseteq H \supseteq SO_A(L)ker\theta_A SO(V),$$

there is a sublattice $N \subseteq L$ of codimension zero such that

$$H = SO(V)X_A(L/N)\ker\theta_A.$$

In particular, the number of spinor genera in $gen(L)$ which represent the given form is a power of 2 and any possible power of 2 can occur.

PROOF. By applying the spinor norm maps on adelic group $SO_A(V)$, we have the following isomorphism

$$\theta_A = \prod_{\mathfrak{p}\in\Omega}\theta_\mathfrak{p} : SO_A(V)/SO(V)SO_A(L)ker\theta_A \cong I_F/\dot{F}\theta_A(SO_A(L)).$$

Then $\theta_A(H)$ can be regarded as a finite dimensional $\mathbb{F}_2$-vector space. Let

$$T = \{\,\mathfrak{p}\in\Omega_F\ :\ L_\mathfrak{p}\text{ is not unimodular}\,\}\cup\{\infty\}.$$

By Chebotarev's Theorem (see [We, Chapter XIII, Thm.12]), there are

$$\mathfrak{q}_1,\cdots,\mathfrak{q}_t \in \Omega_F\setminus T$$

and the ideles

$$\{i_1,\cdots,i_t\}\ \text{as the basis of }\theta_A(H)$$

such that

$$\text{the }\mathfrak{p}\text{-component of }i_k = \begin{cases}\pi_{\mathfrak{q}_k} & \text{if }\mathfrak{p} = \mathfrak{q}_k \\ 1 & \text{otherwise}\end{cases}$$

for $1 \leq k \leq t$. Then N is obtained by

$$N_\mathfrak{p} = \begin{cases}\mathfrak{p}L_\mathfrak{p} & \mathfrak{p} = \mathfrak{q}_k \text{ for } 1 \leq k \leq t \\ L_\mathfrak{p} & \text{otherwise.}\end{cases}$$

By Lemma 1.7.3, one has

$$i_k \in \dot{F}\theta_A(X(L/N))$$

for $1 \leq k \leq t$ and

$$[\dot{F}\theta_A(X(L/N)) : \dot{F}\theta_A(SO_A(L))] = 2^t.$$

Therefore

$$\theta_A(H) = \dot{F}\theta_A(X(L/N)).$$

$\square$

Finally one also can establish a similar result as Thm.2.2.7 to determine if K is represented by $spn(L)$. For any $M \in cls(L)$ satisfying (c22), we can write

$$[K : K\cap M] = \prod_{\mathfrak{p}\notin S}\mathfrak{p}^{d_\mathfrak{p}}.$$

2.3. REPRESENTATIONS OF SPINOR GENERA, CODIMENSION 0

One can define an idele i_M associated to M as follows

$$\text{the } \mathfrak{p}\text{-component of } i_M = \begin{cases} \pi_\mathfrak{p}^{d_\mathfrak{p}} & \text{if } \mathfrak{p} \notin S \\ 1 & \text{otherwise.} \end{cases}$$

Let
$$\Im(L, K) = \{i_M \in I_F \ : \ M \in cls(L) \text{ satisfying } (c22) \ \}.$$

THEOREM 2.3.2. *Suppose K is represented by $gen(L)$ with codimension 0. Then K is represented by $spn(L)$ if and only if there is $i \in \Im(L, K)$ such that*
$$i \in \dot{F}\theta_A(gen(L) : K).$$

PROOF. Let $M \in cls(L)$ satisfying (c22) such that $i = i_M$. Since K is represented by $gen(L)$, there is
$$\tau = (\tau_\mathfrak{p})_{\mathfrak{p} \in \Omega_F} \in SO_A(FL) \quad \text{such that} \quad K \subseteq \tau(M).$$
By (c22), one can assume that
$$\tau_\mathfrak{p} = 1 \quad \text{for } \mathfrak{p} \in S.$$
Then
$$i \in \dot{F}\theta_A(gen(L) : K)$$
$$\Leftrightarrow \theta_A(\tau) \in \dot{F}\theta_A(gen(L) : K)$$
$$\Leftrightarrow K \text{ is represented by } spn(M) = spn(L)$$
by Prop.2.2.5. □

Bibliography

[Ar] Arf, C., Untersuchungen über quadratische Formen in Körpern der Charakteristik 2 (Teil I.), *J.reine angew. Math.* 183 (1940), 148-167.

[Be] Beli, C.N., Integral spinor norm groups over dyadic local fields, *J. Number Theory*.102 (2003) 125-182.

[Ch] Chevalley, C.C., The Algebraic Theory of Spinors, Columbia University, 1954.

[Co] Connors,E.A., Finiteness of class number in characteristic 2,*J. Number Theory.*, 4 (1972) 191-218.

[CX] Chan, W. & Xu, F., On representations of spinor genera, *Compositio Math.*, 140 (2004) 287-300.

[De] Dieudonne,J., La geometrie des groupes classiques, in *Ergebnisse der Mathematik und ihrer Grenzgebiete*, Heft 5, Springer-Verlag, 1955.

[Ei] Eichler, M., Die Ähnlichkeitsklassen indefiniter Gitter, *Math.Z.*, 55 (1952) 216-252.

[Ha] Harder,G.,Minkowskische Reduktionstheorie über Funktionenkörpern, *Invent. Math.*, 7 (1969) 33-54.

[HSX] Hsia,J.S. Shao, Y.Y. & Xu, F., Representations of indefinite quadratic forms, *J. reine angew. Math.*, 494 (1998) 129-140.

[Kn] Kneser,M., Klassenzahlen indefiniter quadratischer Formen in drei oder mehr Veränderlichen, *Arch. Math.*, VII (1956) 323-332.

[Kn1] Kneser,M., Darstellungsmasse indefiniter quadratischer Formen, *Math. Z.* 77 (1961) 188–194.

[Kn2] Kneser, M., Representations of integral quadratic forms, *Canadian Mathematical Society, Conference Proceedings*, 4 (1984) 159-172.

[OP] O'Meara,O.T. & Pollak,B., Generation of local integral orthogonal groups, *Math. Zeit.* 87 (1965) 385-400.

[OM] O'Meara,O.T., The integral representations of quadratic forms over local fields, *Amer. J. Math.* 80 (1958) 843–878.

[OM1] O'Meara,O.T., Introduction to Quadratic Forms, Springer-Verlag, 1973.

[OM2] O'Meara,O.T., Hilbert's eleventh problem: the arithmetic theory of quadratic forms. Mathematical developments arising from Hilbert problems, *Proc. Sympos. Pure Math., Amer. Math. Soc., Providence, R. I.* Vol. XXVIII, (1975) 379-400.

[Po] Pollak, B., Generation of local integral orthogonal groups in characteristic 2,*Canad. J. Math.* 20 (1968) 1178-1191.

[Pr] Prasad, G., Strong approximation for semi-simple groups over function fields, *Ann. of Math.*, 105 (1977) 553-572.

[Ri] Riehm,C.R.,On the integral representations of quadratic forms over local fields, *Amer. J. Math.*, 86 (1964) 25–62.

[Ri1] Riehm,C.R., Integral representations of quadratic forms in characteristic 2, *Amer. J. Math.* 87 (1965) 32-64.

[Sa] Sah, C.H., Quadratic forms over fields of characteristic 2, *Amer. J. Math.* 82 (1960) 812-830.

[Se] Serre, J.P., Corps Locaux, Hermann, Paris, 1962.

[Si] Siegel, C.L., Über die analytische Theorie der quadratischen Formen I, II, III, *Ann. of Math.* 36, 37, 38 (1935) (1936) (1937) 527-606, 230-263, 212-291.

[SP] Schulze-Pillot,R., Darstellung durch Spinorgeschlechter ternärer quadratischer Formen, *J. Number Theory* 12 (1980) 529-540.

[We] Weil, A., Sur la théorie des formes quadratiques, Colloq. Théorie des Groupes Algébriques (1962) 9–22.

[We1] Weil, A., Basic Number Theory, Springer-Verlag, 1974.

[Wi] Witt, E., Theorie der quadratischen Formen in beliebigen Körpern, *J. reine angew. Math.* 176 (1937) 31-44.

[Xu] Xu, F., Spinor norms of local integral rotations in characteristic 2, *J. Number Theory* 32 (1989) 289-296.

[Xu1] Xu, F., Generation of integral orthogonal groups over dyadic local fields, *Pacific J. Math.* 167 (1995) 385-398.

[Xu2] Xu, F., Minimal norm Jordan splittings of quadratic lattices over complete dyadic discrete valuation rings, *Arch. Math.* 81 (2003) 402-415.

[Xu3] Xu, F., On representations of spinor genera, II, *Math.Ann.* 332 (2005) 37-53.

Editorial Information

To be published in the *Memoirs*, a paper must be correct, new, nontrivial, and significant. Further, it must be well written and of interest to a substantial number of mathematicians. Piecemeal results, such as an inconclusive step toward an unproved major theorem or a minor variation on a known result, are in general not acceptable for publication.

Papers appearing in *Memoirs* are generally at least 80 and not more than 200 published pages in length. Papers less than 80 or more than 200 published pages require the approval of the Managing Editor of the Transactions/Memoirs Editorial Board.

As of March 31, 2008, the backlog for this journal was approximately 17 volumes. This estimate is the result of dividing the number of manuscripts for this journal in the Providence office that have not yet gone to the printer on the above date by the average number of monographs per volume over the previous twelve months, reduced by the number of volumes published in four months (the time necessary for preparing a volume for the printer). (There are 6 volumes per year, each usually containing at least 4 numbers.)

A Consent to Publish and Copyright Agreement is required before a paper will be published in the *Memoirs*. After a paper is accepted for publication, the Providence office will send a Consent to Publish and Copyright Agreement to all authors of the paper. By submitting a paper to the *Memoirs*, authors certify that the results have not been submitted to nor are they under consideration for publication by another journal, conference proceedings, or similar publication.

Information for Authors

Memoirs are printed from camera copy fully prepared by the author. This means that the finished book will look exactly like the copy submitted.

Initial submission. The AMS uses Centralized Manuscript Processing for initial submissions. Authors should submit a PDF file using the Initial Manuscript Submission form found at www.ams.org/cgi-bin/peertrack/submission.pl, or send one copy of the manuscript to the following address: Centralized Manuscript Processing, MEMOIRS OF THE AMS, 201 Charles Street, Providence, RI 02904-2294 USA. If a paper copy is being forwarded to the AMS, indicate that it is for it Memoirs and include the name of the corresponding author, contact information such as email address or mailing address, and the name of an appropriate Editor to review the paper (see the list of Editors below).

The paper must contain a *descriptive title* and an *abstract* that summarizes the article in language suitable for workers in the general field (algebra, analysis, etc.). The *descriptive title* should be short, but informative; useless or vague phrases such as "some remarks about" or "concerning" should be avoided. The *abstract* should be at least one complete sentence, and at most 300 words. Included with the footnotes to the paper should be the 2000 *Mathematics Subject Classification* representing the primary and secondary subjects of the article. The classifications are accessible from www.ams.org/msc/. The list of classifications is also available in print starting with the 1999 annual index of *Mathematical Reviews*. The Mathematics Subject Classification footnote may be followed by a list of *key words and phrases* describing the subject matter of the article and taken from it. Journal abbreviations used in bibliographies are listed in the latest *Mathematical Reviews* annual index. The series abbreviations are also accessible from www.ams.org/publications/. To help in preparing and verifying references, the AMS offers MR Lookup, a Reference Tool for Linking, at www.ams.org/mrlookup/.

Electronically prepared manuscripts. The AMS encourages electronically prepared manuscripts, with a strong preference for $\mathcal{A}_{\mathcal{M}}\mathcal{S}$-LaTeX. To this end, the Society has prepared $\mathcal{A}_{\mathcal{M}}\mathcal{S}$-LaTeX author packages for each AMS publication. Author packages include instructions for preparing electronic manuscripts, samples, and a style file that generates the particular design specifications of that publication series. Though $\mathcal{AMS}$-LaTeX is the highly preferred format of TeX, author packages are also available in $\mathcal{AMS}$-TeX.

Authors may retrieve an author package from the AMS website starting from www.ams.org/tex/ or via FTP to ftp.ams.org (login as anonymous, enter username as password, and type cd pub/author-info). The *AMS Author Handbook* and the *Instruction Manual* are available in PDF format following the author packages link from www.ams.org/tex/. The author package can also be obtained free of charge by sending email to tech-support@ams.org (Internet) or from the Publication Division, American Mathematical Society, 201 Charles St., Providence, RI 02904-2294, USA. When requesting an author package, please specify $\mathcal{AMS}$-LaTeX or $\mathcal{AMS}$-TeX and the publication in which your paper will appear. Please be sure to include your complete mailing address.

After acceptance. The final version of the electronic file should be sent to the Providence office (this includes any TeX source file, any graphics files, and the DVI or PostScript file) immediately after the paper has been accepted for publication.

Before sending the source file, be sure you have proofread your paper carefully. The files you send must be the EXACT files used to generate the proof copy that was accepted for publication. For all publications, authors are required to send a printed copy of their paper, which exactly matches the copy approved for publication, along with any graphics that will appear in the paper.

Accepted electronically prepared files can be submitted via the web at www.ams.org/submit-book-journal/, sent via FTP, or sent on CD-Rom or diskette to the Electronic Prepress Department, American Mathematical Society, 201 Charles Street, Providence, RI 02904-2294 USA. TeX source files, DVI files, and PostScript files can be transferred over the Internet by FTP to the Internet node ftp.ams.org (130.44.1.100). When sending a manuscript electronically via CD-Rom or diskette, please be sure to include a message identifying the paper as a Memoir.

Electronically prepared manuscripts can also be sent via email to pub-submit@ams.org (Internet). In order to send files via email, they must be encoded properly. (DVI files are binary and PostScript files tend to be very large.)

Electronic graphics. Comprehensive instructions on preparing graphics are available at www.ams.org/jourhtml/. A few of the major requirements are given here.

Submit files for graphics as EPS (Encapsulated PostScript) files. This includes graphics originated via a graphics application as well as scanned photographs or other computer-generated images. If this is not possible, TIFF files are acceptable as long as they can be opened in Adobe Photoshop or Illustrator. No matter what method was used to produce the graphic, it is necessary to provide a paper copy to the AMS.

Authors using graphics packages for the creation of electronic art should also avoid the use of any lines thinner than 0.5 points in width. Many graphics packages allow the user to specify a "hairline" for a very thin line. Hairlines often look acceptable when proofed on a typical laser printer. However, when produced on a high-resolution laser imagesetter, hairlines become nearly invisible and will be lost entirely in the final printing process.

Screens should be set to values between 15% and 85%. Screens which fall outside of this range are too light or too dark to print correctly. Variations of screens within a graphic should be no less than 10%.

Inquiries. Any inquiries concerning a paper that has been accepted for publication should be sent to memo-query@ams.org or directly to the Electronic Prepress Department, American Mathematical Society, 201 Charles St., Providence, RI 02904-2294 USA.

Editors

This journal is designed particularly for long research papers, normally at least 80 pages in length, and groups of cognate papers in pure and applied mathematics. Papers intended for publication in the *Memoirs* should be addressed to one of the following editors. The AMS uses Centralized Manuscript Processing for initial submissions to AMS journals. Authors should follow instructions listed on the Initial Submission page found at www.ams.org/memo/memosubmit.html.

Algebra to ALEXANDER KLESHCHEV, Department of Mathematics, University of Oregon, Eugene, OR 97403-1222; email: ams@noether.uoregon.edu

Algebraic geometry and its application to MINA TEICHER, Emmy Noether Research Institute for Mathematics, Bar-Ilan University, Ramat-Gan 52900, Israel; email: teicher@macs.biu.ac.il

Algebraic geometry to DAN ABRAMOVICH, Department of Mathematics, Brown University, Box 1917, Providence, RI 02912; email: amsedit@math.brown.edu

Algebraic number theory to V. KUMAR MURTY, Department of Mathematics, University of Toronto, 100 St. George Street, Toronto, ON M5S 1A1, Canada; email: murty@math.toronto.edu

Algebraic topology to ALEJANDRO ADEM, Department of Mathematics, University of British Columbia, Room 121, 1984 Mathematics Road, Vancouver, British Columbia, Canada V6T 1Z2; email: adem@math.ubc.ca

Combinatorics to JOHN R. STEMBRIDGE, Department of Mathematics, University of Michigan, Ann Arbor, Michigan 48109-1109; email: FRS@umich.edu

Complex analysis and harmonic analysis to ALEXANDER NAGEL, Department of Mathematics, University of Wisconsin, 480 Lincoln Drive, Madison, WI 53706-1313; email: nagel@math.wisc.edu

Differential geometry and global analysis to LISA C. JEFFREY, Department of Mathematics, University of Toronto, 100 St. George St., Toronto, ON Canada M5S 3G3; email: jeffrey@math.toronto.edu

Dynamical systems and ergodic theory and complex anaysis to YUNPING JIANG, Department of Mathematics, CUNY Queens College and Graduate Center, 65-30 Kissena Blvd., Flushing, NY 11367; email: Yunping.Jiang@qc.cuny.edu

Functional analysis and operator algebras to DIMITRI SHLYAKHTENKO, Department of Mathematics, University of California, Los Angeles, CA 90095; email: shlyakht@math.ucla.edu

Geometric analysis to WILLIAM P. MINICOZZI II, Department of Mathematics, Johns Hopkins University, 3400 N. Charles St., Baltimore, MD 21218; email: trans@math.jhu.edu

Geometric analysis to MARK FEIGHN, Math Department, Rutgers University, Newark, NJ 07102; email: feighn@andromeda.rutgers.edu

Harmonic analysis, representation theory, and Lie theory to ROBERT J. STANTON, Department of Mathematics, The Ohio State University, 231 West 18th Avenue, Columbus, OH 43210-1174; email: stanton@math.ohio-state.edu

Logic to STEFFEN LEMPP, Department of Mathematics, University of Wisconsin, 480 Lincoln Drive, Madison, Wisconsin 53706-1388; email: lempp@math.wisc.edu

Number theory to JONATHAN ROGAWSKI, Department of Mathematics, University of California, Los Angeles, CA 90095; email: jonr@math.ucla.edu

Partial differential equations to GUSTAVO PONCE, Department of Mathematics, South Hall, Room 6607, University of California, Santa Barbara, CA 93106; email: ponce@math.ucsb.edu

Partial differential equations and dynamical systems to PETER POLACIK, School of Mathematics, University of Minnesota, Minneapolis, MN 55455; email: polacik@math.umn.edu

Probability and statistics to RICHARD BASS, Department of Mathematics, University of Connecticut, Storrs, CT 06269-3009; email: bass@math.uconn.edu

Real analysis and partial differential equations to DANIEL TATARU, Department of Mathematics, University of California, Berkeley, Berkeley, CA 94720; email: tataru@math.berkeley.edu

All other communications to the editors should be addressed to the Managing Editor, ROBERT GURALNICK, Department of Mathematics, University of Southern California, Los Angeles, CA 90089-1113; email: guralnic@math.usc.edu.

Titles in This Series

909 **Cameron McA. Gordon and Ying-Qing Wu,** Toroidal Dehn fillings on hyperbolic 3-manifolds, 2008

908 **J.-L. Waldspurger,** L'endoscopie tordue n'est pas si tordue, 2008

907 **Yuanhua Wang and Fei Xu,** Spinor genera in characteristic 2, 2008

906 **Raphaël S. Ponge,** Heisenberg calculus and spectral theory of hypoelliptic operators on Heisenberg manifolds, 2008

905 **Dominic Verity,** Complicial sets characterising the simplicial nerves of strict ω-categories, 2008

904 **William M. Goldman and Eugene Z. Xia,** Rank one Higgs bundles and representations of fundamental groups of Riemann surfaces, 2008

903 **Gail Letzter,** Invariant differential operators for quantum symmetric spaces, 2008

902 **Bertrand Toën and Gabriele Vezzosi,** Homotopical algebraic geometry II: Geometric stacks and applications, 2008

901 **Ron Donagi and Tony Pantev (with an appendix by Dmitry Arinkin),** Torus fibrations, gerbes, and duality, 2008

900 **Wolfgang Bertram,** Differential geometry, Lie groups and symmetric spaces over general base fields and rings, 2008

899 **Piotr Hajłasz, Tadeusz Iwaniec, Jan Malý, and Jani Onninen,** Weakly differentiable mappings between manifolds, 2008

898 **John Rognes,** Galois extensions of structured ring spectra/Stably dualizable groups, 2008

897 **Michael I. Ganzburg,** Limit theorems of polynomial approximation with exponential weights, 2008

896 **Michael Kapovich, Bernhard Leeb, and John J. Millson,** The generalized triangle inequalities in symmetric spaces and buildings with applications to algebra, 2008

895 **Steffen Roch,** Finite sections of band-dominated operators, 2008

894 **Martin Dindoš,** Hardy spaces and potential theory on C^1 domains in Riemannian manifolds, 2008

893 **Tadeusz Iwaniec and Gaven Martin,** The Beltrami Equation, 2008

892 **Jim Agler, John Harland, and Benjamin J. Raphael,** Classical function theory, operator dilation theory, and machine computation on multiply-connected domains, 2008

891 **John H. Hubbard and Peter Papadopol,** Newton's method applied to two quadratic equations in $\mathbb{C}^2$ viewed as a global dynamical system, 2008

890 **Steven Dale Cutkosky,** Toroidalization of dominant morphisms of 3-folds, 2007

889 **Michael Sever,** Distribution solutions of nonlinear systems of conservation laws, 2007

888 **Roger Chalkley,** Basic global relative invariants for nonlinear differential equations, 2007

887 **Charlotte Wahl,** Noncommutative Maslov index and eta-forms, 2007

886 **Robert M. Guralnick and John Shareshian,** Symmetric and alternating groups as monodromy groups of Riemann surfaces I: Generic covers and covers with many branch points, 2007

885 **Jae Choon Cha,** The structure of the rational concordance group of knots, 2007

884 **Dan Haran, Moshe Jarden, and Florian Pop,** Projective group structures as absolute Galois structures with block approximation, 2007

883 **Apostolos Beligiannis and Idun Reiten,** Homological and homotopical aspects of torsion theories, 2007

882 **Lars Inge Hedberg and Yuri Netrusov,** An axiomatic approach to function spaces, spec tral synthesis and Luzin approximation, 2007

881 **Tao Mei,** Operator valued Hardy spaces, 2007

TITLES IN THIS SERIES

880 **Bruce C. Berndt, Geumlan Choi, Youn-Seo Choi, Heekyoung Hahn, Boon Pin Yeap, Ae Ja Yee, Hamza Yesilyurt, and Jinhee Yi,** Ramanujan's forty identities for Rogers-Ramanujan functions, 2007
879 **O. García-Prada, P. B. Gothen, and V. Muñoz,** Betti numbers of the moduli space of rank 3 parabolic Higgs bundles, 2007
878 **Alessandra Celletti and Luigi Chierchia,** KAM stability and celestial mechanics, 2007
877 **María J. Carro, José A. Raposo, and Javier Soria,** Recent developments in the theory of Lorentz spaces and weighted inequalities, 2007
876 **Gabriel Debs and Jean Saint Raymond,** Borel liftings of Borel sets: Some decidable and undecidable statements, 2007
875 **C. Krattenthaler and T. Rivoal,** Hypergéométrie et fonction zêta de Riemann, 2007
874 **Sonia Natale,** Semisolvability of semisimple Hopf algebras of low dimension, 2007
873 **A. J. Duncan,** Exponential genus problems in one-relator products of groups, 2007
872 **Anthony V. Geramita, Tadahito Harima, Juan C. Migliore, and Yong Su Shin,** The Hilbert function of a level algebra, 2007
871 **Pascal Auscher,** On necessary and sufficient conditions for L^p-estimates of Riesz transforms associated to elliptic operators on $\mathbb{R}^n$ and related estimates, 2007
870 **Takuro Mochizuki,** Asymptotic behaviour of tame harmonic bundles and an application to pure twistor D-modules, Part 2, 2007
869 **Takuro Mochizuki,** Asymptotic behaviour of tame harmonic bundles and an application to pure twistor D-modules, Part 1, 2007
868 **Gelu Popescu,** Entropy and multivariable interpolation, 2006
867 **Vilmos Totik,** Metric properties of harmonic measures, 2006
866 **William Craig,** Semigroups underlying first-order logic, 2006
865 **Nathanial P. Brown,** Invariant means and finite representation theory of $C*$-algebras, 2006
864 **John M. Lee,** Fredholm operators and Einstein metrics on conformally compact manifolds, 2006
863 **M. Lübke and A. Teleman,** The Universal Kobayashi-Hitchin correspondence on Hermitian manifolds, 2006
862 **Alberto Canonaco,** The Beilinson complex and canonical rings of irregular surfaces, 2006
861 **Leon A. Takhtajan and Lee-Peng Teo,** Weil-Petersson metric on the universal Teichmüller space, 2006
860 **Thomas M. Fiore,** Pseudo limits, biadjoints and pseudo algebras: Categorical foundations of conformal field theory, 2006
859 **N. Arcozzi, R. Rochberg, and E. Sawyer,** Carleson measures and interpolating sequences for Besov spaces on complex balls, 2006
858 **Enrico Valdinoci, Berardino Sciunzi, and Vasile Ovidiu Savin,** Flat level set regularity of p-Laplace phase transitions, 2006
857 **Donatella Danielli, Nocola Garofalo, and Duy-Minh Nhieu,** Non-doubling Ahlfors measures, perimeter measures, and the characterization of the trace spaces of Sobolev functions in Carnot-Carathéodory spaces, 2006
856 **Vladimir Bolotnikov and Harry Dym,** On boundary interpolation for matrix valued Schur functions, 2006
855 **Yevgenia Kashina, Yorck Sommerhäuser, and Yongchang Zhu,** On higher Frobenius-Schur indicators, 2006

For a complete list of titles in this series, visit the
AMS Bookstore at **www.ams.org/bookstore/**.